# BUILDING ON THE SEA

# BUILDING ON THE SEA

## FORM AND MEANING IN MODERN SHIP ARCHITECTURE

### PETER QUARTERMAINE

AD ACADEMY EDITIONS    NATIONAL MARITIME MUSEUM

**Dedication: For Tony, casting off 1996**

FOREWORD

The claim that Britain is a maritime nation is a familiar cliche but one which is no longer true. Britain is certainly still an island and over ninety per cent of its exports and imports are carried by ships, but today these ships are not built in Britain. Ships play an equally vital part in global economics, perhaps more so now than ever, but they have disappeared from our visual landscape. The world's largest moving man-made objects travel daily across our seas but they operate in a forgotten space. Ships have vanished from the hearts of our busy cities; they appear no more on the horizon of the public mind than they do in the public eye.

This book is concerned with the cultural and imaginative significance of the modern ship, its form and meaning. In an age when at first glance the design of cargo vessels appears to be prosaic to the point of ugliness, and that of cruise liners appears to spring from Hollywood film studios, it seeks to reclaim ships as architecture.

Ships are rarely discussed outside maritime circles. If we want to ensure that they live on in the public mind beyond the historiography of the heritage industry then the importance of books such as this cannot be underestimated.

Richard Ormond, Director, National Maritime Museum

AUTHOR'S NOTE

This text was researched and written in non-existent spare time. As to motive I can only modestly echo Björn Landström in his Preface to *The Ship* over thirty years ago: 'This book is an attempt to satisfy a need. I had hoped for many years that such a book existed, and had searched for it on the shelves of bookshops and public libraries.'

Although *Building on the Sea* addresses a field complementary to my professional work in post-colonial media and the built environment, it addresses topics that have rarely been out of my mind since the late 1950s. It is written by an amateur (in both senses) but the positive response to this project from industry internationally, as well as from colleagues worldwide in photography, architecture and design, has confirmed the importance of ships. I hope that this book may stimulate further work which will trace the commercial and design significance of modern ships, passenger and cargo alike, and that I can help with such work. It is encouraging that both shipbuilders and shipowners hold design in high regard, for sound professional and commercial reasons. David Pye's fine 1950 study recognised just such commercial priorities and my own book is indebted to that dedicated professionalism which builds on the sea structures of both extraordinary beauty and outstanding efficiency.

Specially commissioned photography, 1995, by Tina Chambers and Peter Robinson, ©National Maritime Museum, Greenwich.

FRONT COVER: *Duc de Normandie*, Britanny Ferries
BACK COVER: Container vessels at Tilbury Dock – foreground: *Alekander Prokofyev*, background: *Matumba*
PAGE 2: *Tempere*, container vessel at Tilbury Dock

First published in Great Britain in 1996 by
ACADEMY EDITIONS
An imprint of

ACADEMY GROUP LTD
42 Leinster Gardens, London W2 3AN
Member of the VCH Publishing Group

ISBN 1 85490 446 9

Distributed to the trade in the United States of America by
NATIONAL BOOK NETWORK, INC
4720 Boston Way, Lanham, Maryland 20706

Printed and bound in the United Kingdom

# CONTENTS

# BEYOND NOSTALGIA
## APPRECIATING MODERN SHIPS

*The beauty of ships can be appreciated without technical knowledge of shipbuilding and the sea, just as the countryside can be appreciated without an understanding of geology and agriculture. It would be a pity if we shut our eyes to everything we did not understand.*

David Pye, The Things We See: Ships, *1950*

*What is the essence of a building? Well, it must protect you from the elements, keep you warm when it's cold, cool when it's hot, and it should tell you something about its age – its place in time.*

Sir Norman Foster, 'Boeing 747' in Building Sights, *1995*

The very title of this book is problematic since ships are not usually discussed as architecture; indeed, rarely discussed at all. Yet David Pye's contention that the overriding and beneficial concern of ship designers has been with efficiency rather than 'style', a comment from a distinguished designer and teacher who also wrote a fine early study of ship design, immediately questions the relative influence of function and aesthetics in their design, a debate complicated by the fact that traditionally naval architects have exercised their very considerable skills in a reserved manner which contrasts markedly with the attention-seeking public profiles adopted by some distinguished (and less distinguished) architects.[1] Because ships are rarely the subject of public debate (though the eventual fate of the Royal Yacht *Britannia* may prove an interesting exception) even the Prince of Wales has kept aloof, and ship designers have been able to exercise their skills with a long-established reticence in which they take some pride, even though the design, function and technology of ships themselves has changed out of all recognition in the last few decades. Such individuals take both their capabilities and their responsibilities (for which they must also carry massive indemnity insurance) very seriously; we should pray not only for those who go to sea in ships, but also for those who design them. Naval architects have made modesty and invisibility a point of honour, and even now it is not easy to identify them. Much of the essential design work on a ship – even a new kind of ship – draws upon a wealth of acquired and inherited experience for the simple reason that the demands of the sea and wind are largely unchanging, at least within any particular geographical area, and it therefore makes good sense to place trust in the wisdom of years. The same does not apply to technology, either in propulsion or building, in which there have been spectacular developments, including some which led in the postwar period to the demise of the British shipbuilding industry.

Fantasy still plays a significant role in our attitudes to ships and the sea but those who design ships must be severely practical if their work is to stay afloat

*OPPOSITE: 'Heroic outer form' – the launch of the aircraft carrier* HMS Victorious *at the Vickers-Armstrong yard on the Tyne, 14 September 1939; FROM ABOVE: Ship architecture in a now rare rural context – the coaster* Adi *leaves Exeter for the sea in 1963; A deckchair view of modern shipping – the* Celtic Navigator *leaves Teignmouth, summer 1995*

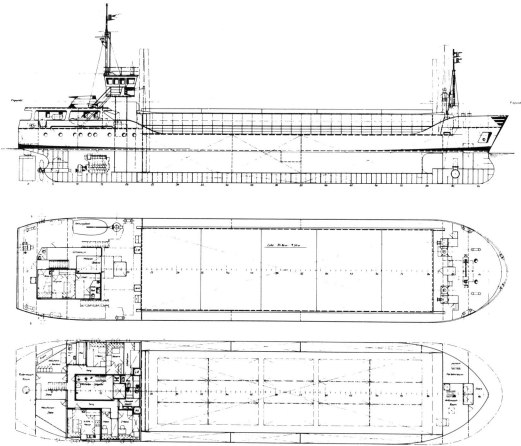

profitably. There is little place in marine design and engineering for some of the more whimsical theories, let alone practicalities, of land architecture, and if this is undoubtedly a restriction it is also one within which I am happy to work for the purposes of this book. Practical restrictions have rarely prevented the creation of aesthetically satisfying objects, and of this principle ships remain an important and absorbing illustration.

My own city of Exeter was a working port for several hundred years until the 1970s, when its harbour basin became a maritime museum and yacht mooring; the river ports of Topsham and Exmouth have also since died, the latter only within the past few years. This, however, has always been a pattern and to walk the massive stone quays in any of the mostly-abandoned ports of earlier centuries is a sobering experience. So much effort and skill was expended in outwitting time and tide to fashion a shelter for ships, and now it is all in decay – at best deemed part of 'the heritage' and subject to the often doubtful benefits of 'preservation'. Nowhere is original skill and effort more apparent than in the small ports of the Scottish North-East coast, where terrain and weather alike conspired to make every harbour a challenge built in stone, and where most such ports are now little used.[2] On another scale, the original pattern of London's docklands is hardly matched by the mostly tawdry units that have been scattered across their flattened and filled site, and where whole docks have become the watery courtyard for a housing development, as in Wapping, the scale and the massive precision of the stone basins themselves still make the houses seem irrelevant.[3]

The increasing separation between shipping activities and the wider commercial and social life of port cities may be another reason for the comparative neglect of ship architecture, with the one exception of transatlantic liners.[4] On the other hand, ships are often seen as mere pieces of floating technology, impressive in their way but not worthy of the serious debate accorded to 'real' architecture, despite their scale, function and attention to detail. That ships have not received due attention, and so have not accumulated a body of informed aesthetic appreciation outside specialist journals (themselves mostly concerned with technical and safety matters) perpetuates the neglect. The irony is, as this book argues, that many comparisons might follow from recognition of the extent to which modern ships are one element of an increasingly integrated world transport and leisure system: the omnipresent container makes the former clear to us all, while on the leisure front airlines and cruise companies discuss mergers and takeovers. In both cases – the container vessel and the cruise ship – technology is used to the maximum to serve a specific purpose, but that is true of any efficient architectural structure today, from a railway station to an incinerator. Sir Norman Foster's comments on the Boeing 747 apply equally to ships:

> The fact that we call this an aeroplane rather than a building – or engineering rather than architecture – is really a historical hangover because . . . so much of what we have here is genuinely architectural both in its design and its thinking.[5]

His appreciative position – that 'this machine blurs the distinction between technology and building, and what's more it flies!' – would, if applied imaginatively to ships, not only bring a richer understanding of these complex and often beautiful structures (and what's more, they float!) but also correct a somewhat dated vision

*OPPOSITE, FROM ABOVE: Ship architecture as part of the city – the tanker* Siratus *and the western face of the city of Sydney; The 1988 Hammann and Prahm cargo vessel* Walter Hammann, *showing telescopic bridge arrangement; FROM ABOVE: A mostly lost role of the small ship – the coastal tanker* Guidesman *navigates inland from the Exe estuary to Exeter via the ship canal, 1963; The perils of navigation –* Calshot Spit *lightvessel at anchor in The Solent, 1960; The perils of preservation –* Calshot Spit *lightvessel set in concrete, Southampton, 1995*

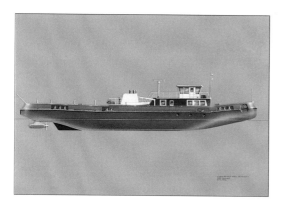

*FROM ABOVE: Indoor construction of a modern passenger vessel at Kvaerner Masa-Yards, Finland; Indoor construction of a modern cargo vessel at Kvaerner Masa-Yards, Finland; A contrast in profiles, visible and invisible – a river icebreaker from Kvaerner Masa-Yards propelled by twin swivelling 'Azipod' units. These are electrically driven from the main engine, which can therefore be located more flexibly within the hullspace than when a propeller shaft is required; OPPOSITE, FROM ABOVE: A new tradition of architecture at sea – the 1992 SWATH cruise liner Radisson Diamond; A shape of the future? Impression of a smooth-profile transatlantic superliner projected by Kvaerner Masa-Yards, Finland*

of ship design on the part of some architects. Foster suggests an understanding of the 747 as 'the ultimate technological building site', a description which certainly fits a modern shipbuilding yard.

Foster's quotation at the head of this chapter differs in emphasis from Pye's but both designers agree that there will be an element in good design which cannot be reduced to mere function – although it should not, and need not, interfere with function. Conversely, some ships must be efficient without much regard to appearance (though all will be as sleek as possible below the waterline). We shall see this principle exemplified in the design of the liner *Queen Elizabeth 2* (or the *QE2* as she is usually known) later in this book, but Foster rightly asserts design as a conscious principle:

There is, I believe, a common misconception about architecture and design – the belief that if the forces of nature are allowed to create form then that form will be automatically beautiful (the 'if it looks right it is right' sort of argument). Personally, I think this is nonsense. There is no doubt that an aircraft is an extreme example, but I cannot believe mere aerodynamics gave this piece of industrial architecture its heroic outer form. This thing was designed. In fact an engineer called Joe Sutter is credited as the chief designer.[6]

Architecture is inescapably utopian – 'heroic' is Foster's term – shaping something out of nothing, and of this impulse the building, launching and navigation of a ship is arguably the finest expression. One aspect of this perfection is elemental opposition: a ship operates on the surface of oceans which cover much of our world and yet which, until recently, seemed unchanging and unchangeable. The ocean offers a stormy challenge to our endeavours while remaining at once unsettling and reassuring. For land architecture water is a problem to be repelled or a decorative element to be played with, but for a ship it is the life element, even though each vessel must be designed and constructed to displace, repel and survive it. It is this seemingly limitless bounty of the sea, uncontrollable in its scope, that Byron invokes in *Childe Harold's Pilgrimage*:

Roll on, thou deep and dark blue Ocean – roll!
Ten thousand fleets sweep over thee in vain;
Man marks the earth with his ruin – his control
Stops with the shore.[7]

Developments in land architecture have been accepted as natural over the centuries, yet our own time so lacks confidence that buildings which replicate an earlier period are preferred to modern architecture. There is, however, a case for saying that the 'Roman' architecture of Caesars Palace in Las Vegas, an architecture which confidently and successfully translates Roman motifs for well-defined commercial purposes, is more honest than Quinlan Terry's supposedly tasteful replication of 'classical' architecture on the Thames at Richmond. There is nothing like Las Vegas in its voracious translation of other architectural styles for its own commercial ends, whereas Terry's facades merely borrow whatever interest and supposed respectability they have.

For the purposes of this study, the chief significance of Las Vegas lies in little-noticed but significant parallels to be drawn between its isolated, gambling-orientated development and certain trends in the design of some American-owned

FROM ABOVE: Pleasure dome – the well-padded First Class
lounge on the 1913 Canadian-Pacific liner Empress of Asia;
Tourism afloat – the 'Passage to India' lounge on the
Carnival Cruises liner Fascination; Model of the cruise liner
Disney Magic, now building at Fincantieri's Venice yard
and due to enter service in January 1998. The ship will
feature distinctive 'traditional' liner elements and distinctively original styling of bow and stern. A second vessel
follows in November 1998

cruise ships. More broadly, there is much to be gained from applying to ships the
principles first advocated in 1972 by Robert Venturi, Denise Scott Brown and
Steven Izenour in *Learning from Las Vegas*, and reaffirmed in their foreword to
Alan Hess's 1993 study, *Viva Las Vegas! After Hours Architecture*:

> large-scale, complex physical environments like Las Vegas should be
> examined open-mindedly, and from political, economic, symbolic, and
> historical as well as formal perspectives.[8]

Such an approach is essential for understanding ship architecture, and especially some recent passenger liners which give opportunities to construct on a
scale, and for a clientele that has always both justified and required the literal
replication of land architecture as part of interior decor. As Sir Hugh Casson
observed in 1969, in a special issue of the *Architectural Review* devoted to the
new *QE2*:

> The dressing-box of history was ruthlessly pillaged, and to walk through the
> public rooms of the great passenger liners of this period was to leaf through
> a gruesome catalogue of architectural styles. In the early twenties new facilities
> were constantly being demanded and the designers responded with a
> will . . . Pompeiian swimming pools . . . Byzantine chapels . . . Viennese
> cafes . . . Rococo cinemas.[9]

Decor at sea can still be riotously pleasurable and overripe, although there has
been a longstanding debate between those designers who believe in the appro-
priateness for ships of 'the nautical style' and those who accept that passengers
like interiors which offer at sea all the reassuring comforts of land. The solution is
not easy, and Casson notes that 'over-shippy' attitudes to interior decor afloat
'can be as romantically unsuitable as plaster swags and wrought-iron grilles'.[10]
As in hotels, the other homes-from-home, a certain excess seems necessary to
reassure passengers that although they booked a voyage they are still at home:

> This need for reassurance touched with illusion has not disappeared even
> today – if expressed in less frankly a theatrical manner. The smoking room of
> a modern liner may look more like a Californian country club than say a
> Warwickshire manor, but the required atmosphere of being for a time in
> another world is still essential.[11]

The contemporary Italian architect Donato Riccesi shows refreshing frankness in
observing that the sumptuous interior decor on the great Italian liners *Rex* and
*Conte di Savoia*, vessels which between the wars carried both the blessing and
the values of Mussolini on the Atlantic, resembled nothing so much as 'the best
brothels of New Orleans'.[12] Some cruise liners today are promoted as 'love ships',
hinting discreetly at one element of their intended functionality, and as Paul
Fussell noted in 1980, detecting a decline from travel to tourism:

> ships have been replaced by cruise ships, small moveable pseudo-places
> making an endless transit between larger fixed pseudo-places. But even a
> cruise ship is preferable to a plane. It is healthier because you can exercise
> on it, and it is more romantic because you can copulate on it.[13]

Mindful of the rather different needs of its future passengers, Disney's new liners
will apparently 'provide completely separate facilities for honeymooners, adults
travelling without children and seniors'.[14]

Those architectural settings in which ship and land architecture are still juxtaposed offer unique opportunities for comparison and contrast, although these have been little pursued:

> Ships suggest mobility; cities, the fixed and immobile structures of civilization. Ships disperse goods and people; cities concentrate them. The profile of the ship is long and horizontal; that of the city tall and vertical. These contrasts can be reversed: the port is where ships are at rest; the city where the pace of life quickens.[15]

Reyner Banham argued the architectural significance of container ports as early as 1969, noting that 'when you get to Tilbury . . . you see little to recall the typical imagery of ports. What you see, more than anything else, is acreage of flat tarmac, or concrete'. He also felt, however, that while it was 'one of the great sights of London', architects would ignore it:

> What are architects going to do with situations like this? As a profession they claim the right and duty to design 'the complete human environment', but one thing they cannot bear to contemplate is a large flat area of anything at all; they whimper in their campari-soda about airports, supermarkets, 'prairie planning in the new towns' and – above all – car parks.[16]

Despite the importance to world trade of the Suez and Panama canals, few of us readily think of the sea as linking continents, more as a barrier to be bridged or flown over; our imaginations are essentially landlocked. However, that liberating sense that in principle one can sail from any one port in the world to another still operates powerfully, if subconsciously, and in this envisioning of the sea as offering possibilities, ships themselves are a part:

> their transient presence on the world's horizons always seems to be like an assurance of companionship – one does not have to be stranded on a desert island to feel that if a ship is in sight all cannot be lost.[17]

Good friends currently in Penang seem closer gazing out to sea from a Devon beach than they do from my office only twenty minutes away in Exeter. This is not all fancy, for the sea remains the most important means of transporting mass between continents – sometimes even within continents and countries themselves. Air travel is faster, but as a mode of travel it lacks both the immediacy and the continuity of water; the waves that lap the shore, and in which we can swim, are the frayed edges of a liquid carpet that really can transport us to the ends of the world. Many of our most basic materials for building, manufacturing, heating and eating arrive from distant ports by ship.

However, unless we live in a modern port we are unlikely today to witness the seaborne traffic which day and night serves the needs of nations and communities worldwide, 'needs' which include orange juice in bulk by tanker from South America for winter breakfast tables in Europe, and live sheep exported from Freemantle in Western Australia for halal slaughter on arrival in the Arabian Gulf. The functions, and hence the forms, of modern ships are many and specialised, and few landlubbers are aware either of this variety or of the extent to which vessels have changed over the last three decades at a pace previously unknown. They may have noticed, however, that the buildings in what were once working ports have increasingly become residential apartments, and the ports themselves locked and tinkling yacht marinas which are dead except during summer weekends; few ports successfully combine history and commerce. The ships have

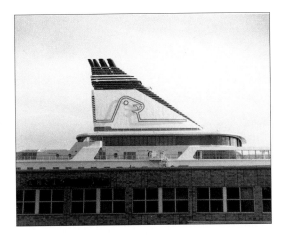

*FROM ABOVE: Integrated transport – the funnel and related superstructure of the superferry Silja Serenade project above her terminal building, Helsinki, 1995; Land link – the Ben Line container ship City of Edinburgh alongside North Port, Port Kelang, Malaysia, c1974; Sea link – the Evergreen International Corporation's container vessel Ever Garland passing through the Panama Canal in 1995*

mostly moved elsewhere, and in their place we are offered a sanitised version of 'maritime heritage':

> The metropolitan gaze no longer falls on the waterfront, and cognitive blankness follows . . . the sea is in many respects less comprehensible to today's elites than it was before 1945, in the nineteenth century, or even during the Enlightenment.[18]

In an article on New York's South Street area, M Christine Boyer, noting that there is 'no better stage set for the spectacle of capital than a recycled mercantile past', details the cultural and historical fascination of such maritime heritage:

> Every waterfront tableau, of course, is in some form or another a seafaring tale that places man in a contest against nature, and every seaport festival marketplace is an entrepôt of exotica, an image theater that reflects our desire for the curious and the marvelous. It is this environment of trade and voyages, of mechanical pulleys, levers, and winches, a world charted by optical instruments and balanced by mechanical laws, that is set up before our eyes in the historic tableau of South Street Seaport.[19]

This book illustrates some of the special problems faced by those who design, build and operate modern ships. It explains these vessels' varied appearance as an aspect of function but, in taking account of the functionalist notion that efficiency makes for good looks, also addresses more complex definitions of overall 'efficiency'. If a liner's additional dummy funnel impresses potential passengers it is efficient, not in dispersing smoke but in attracting passengers. There are analogies here with car design: fins, chrome, sports wheels – even an Italian-sounding 'expresso' name – do not make a car go faster, but if they sell cars they serve their purpose.[20]

Most people are relatively unfamiliar with the range and appearance of many modern ships and this gives visual information special importance in this study. High quality illustration is an integral part of its 'meaning' and photographs and plans are complementary to the text, not a mere adjunct to it, illustrating as they do essential aspects of ship design. Early engineering drawings may indeed, as Francis Klingender has argued, be among the major artifacts of their period, but recognition of the skills they required and fostered – as well as their inherent beauty and practicality – should not blind us to the new and exciting capabilities of the computer.[21] One practical consequence of modern computer assisted design (CAD) techniques is that ships will be easier to upgrade, refit or even convert because their basic design plans are on computer; this may in turn fundamentally change the economics of re-using ships in the future.

Accepting illustrations as integral to a book's intention requires awareness of the relationship between the visual image and the printed word, a field of communications explored by Charles Ivins Jnr and Estelle Jussim.[22] It also requires some curiosity on the part of the reader as to the logic and origins of illustrations, for neither text nor illustration can be innocent. Most colour plates here are selected from photographs specially taken on location after detailed discussion between myself and the photographers themselves, while others were chosen from a range of mostly professional sources in several countries. Monochrome images come either from the National Maritime Museum's archives or from other photographers

*FROM ABOVE: A rare example of old and new still in working harmony, Stockholm waterfront in 1995; Funnel power – the 1913 White Star liner* Aquitania; *OPPOSITE, FROM ABOVE, The 1969* QE2 *(with later staterooms behind the bridge and altered funnel) against the New York skyline in 1989; Italian architectures contrast – the 1993 Costa Cruises liner* Costa Romantica *passes La Salute and Piazza San Marco, Venice*

and collections worldwide, including my own. In using this material there has been continuity of intention between the printed word and the visual image in the book's production, with photographs reinforcing arguments in a text itself shaped in response to photographic images.

Through pictures in museums and galleries, and through images in travel brochures and advertising, painting and photography have historically shaped the conventions by which we see ships. There is no scholarly study of the photographic depiction of ships comparable in scope to Robinson and Herschman's *Architecture Transformed: A History of the Photography of Buildings from 1839 to the Present*, a book its authors describe as 'a history of style in photography', and which argues that 'from the beginning photographers have used their pictures to make critical and expressive statements'.[23] Nevertheless, many of these insights could usefully be applied to maritime topics:

> The manner in which a photographer can approach a building has always been constrained by the nature of the subject and its surroundings . . . Yet an age may find itself especially drawn to a particular kind of view, to one perspective, and pictures of that sort will carry special conviction and deserve attention.[24]

A further complication in understanding the shifting depiction of ships is that their role in visual imagery is complex. In the early decades of this century New York 'was probably depicted more often than any other port' but 'ships appeared more for visual effect than out of a conviction that they and the city belonged together'.[25] Current advertising for cruise ships similarly uses spectacular settings as a location for photographing new vessels, but without any intrinsic architectural interest being posited between ship and shore; the ship itself is exotic.

Marine painting is better served by the critics, but a general aesthetic recoil from modern ship technology, together with changes in marine painting itself, have produced a neglect of the modern period:

> once the naturalism that had, in one guise or another, informed all the developments of marine painting since the eighteenth century, ceased to function as the ground of advanced painters, a gap such as had never before existed opened between the marine artist and those at the forefront of landscape art. The gap would not be closed.[26]

The postwar use of aerial photography, itself a development from military reconnaissance, has changed the way ships are commonly depicted for commercial purposes, though some cruise companies have adopted a self-consciously 'period' style in their publicity, seeking to evoke the elegance and privilege of past eras.[27] Aerial photography has many advantages, and is especially useful for revealing the deck layout and gear of a vessel. From a small plane flying at deck height – with large vessels often lower – the traditional three-quarter bow shot can produce dramatic views of vessels under way. The camera lens also exaggerates perspective, and the low-level bow shot has become a desirable one for builders and owners, combining informative detail with an emphasis on speed and power. Such photographs also convey a strong sense of mass and mobility, with the camera angle usually enabling any background information to be excluded, and these images can be powerful depictions of the sheer scale of some modern

*FROM ABOVE: P&O's* Canberra *passes the finger wharves at New York. The liner was completed in April 1961 and this picture was probably taken shortly afterwards; A very private beach – the pool deck area on the Carnival Cruise Lines'* 1990 Fantasy vessels. Informative though they can be, aerial photographs are mostly avoided in this book for the simple reason that we do not often see ships from the air; the view here is of the crew, passenger or dockside worker rather than that of the seagull.[28]

Unlike other large land architecture constructions, ships are completely self-sufficient entities and move under their own power at considerable speed for their size; a loaded supertanker can displace several hundred thousand tons, travel at the equivalent of almost twenty miles per hour, and take three miles or more to lose way – certainly a more appropriate term in such circumstances than 'stop'. The design of ships is also clearly directional in both function and appearance: even an observer unfamiliar with any particular vessel could probably tell which way it was intended to travel.

The need for the hull to present as little resistance as practicable (bearing in mind capacity, economy and stability) gives a form which is symmetrical in transverse section but varies greatly in longitudinal section. As with all solid forms which penetrate a medium at anything over minimal speed, the underwater form of a ship's hull needs symmetry for directional balance; even above the waterline only comparatively recent commercial specialisations have brought a break with traditionally 'balanced' designs. The increasing need to accommodate modern onboard technology, whether sophisticated cargo-handling cranes and gantries, or roll-on roll-off vehicle facilities, has had a decisive impact on the design and appearance of modern cargo vessels. The design of passenger vessels reflects different priorities, as we shall see later.

Our response to ships, even as casual observers, is strongly influenced by our attributing to them symbolic characteristics, a response also evoked by other architectural forms: the inviting porch or overhanging eaves of a house, the reassuring sweep or solid mass of a bridge, the forbidding walls of a castle. Despite this, few commentators have addressed this significant element of form:

> While architects have adapted the simple forms of vernacular architecture, they have largely ignored the complex symbolism behind them . . . have discounted the symbolic values that invest these forms and dominate, so anthropologists tell us, the artifactual environment of primitive cultures, often contradicting function and structure in their influence on form.[29]

Ships greatly accentuate such a response by their apparently free-ranging mobility, which is quite different from the constrained movement of a train or car, and more understandable in its modest speed than an aeroplane. Casson commented in 1969:

> Motor cars, once cherished, groomed and regularly exercised as horses, are now as anonymous as milk-bottles. Trains are still admired from the platform but treated by the voyager, once aboard, as shabbily as litter bins. Aircraft have kept the power to awe and even to alarm but seldom to inspire love. Hovercraft with their capricious skirts arouse curiosity and little else. Ships alone invite and receive an affection that is almost personal in its intensity.[30]

This affection encourages symbolic and metaphorical interpretations, and is most obvious in the feminine gendering of ships by the overwhelmingly male maritime community, whether seafarers, academics or fiction writers. Many such common

*FROM ABOVE: Directionality in action – the 1965 traditional cargo vessel* Norma *en route from Hamburg to Helsinki; The 1993* Costa Romantica *demonstrates clean symmetry of deck layout*

symbols and attributions have been reduced to little more than cliches by over-use and this study suggests some basic principles by which ships may be under-stood, not only aesthetically but as intricately functioning machines; also, in some cases, as objects which, as Alan Colquhoun has argued, were 'made ostensibly with utilitarian purposes in mind', yet 'quickly became gestalt entities . . . imbued with aesthetic unity'.[31]

As with land architecture, much of the interest and satisfaction of ship architec-ture comes from understanding a visual design 'language' shared with variations by many vessels over the years, as well as from contrasts of scale, profile and function. Such understanding is greatly enhanced by seeing such contrasts not as contrived for effect but as essential to a vessel's function. In one sense a ship is best seen laden and underway at sea, for then it is functioning as intended and also conforming visually, and in its entirety, to design principles of proportion, elevation and paintwork which can change dramatically when a vessel is in bal-last. This can also cause navigation problems, especially on some smaller ves-sels which have a low air draught for passing under river bridges to inland ports, and are hence fitted with hydraulically-operated telescopic bridges. There are also real virtues in looking closely at ship design in detail, and so the majority of photographs here are taken to provide insight into the functioning of the varied range of equipment with which modern vessels are equipped.

Many of the basic constraints on ship design apply to vessels large and small, but scale can define important aspects of both function and appearance, not least because modern technology has made it feasible to operate very large vessels with quite small crews, and so superstructure often assumes less of a role than it did in earlier periods. One consideration which follows from the directional nature of a ship's overall design – a characteristic not necessarily present in its separate elements – is the relation of the bridge to the bulk of the entire vessel. The bridge and its wheelhouse traditionally formed the nerve centre from which a vessel was steered and from which engines were controlled by telegraph signals. On modern vessels, however, the bridge is often completely enclosed because of increasingly sophisticated electronic equipment, and any distinction between 'bridge' and wheelhouse lost; engines and other equipment are now controlled directly from the bridge.[32] Excellent visibility is always a prime requirement, and for modern vessels not only of the route ahead; design increasingly reflects the requirement for all-round visibility, and one effect has been to offset funnels to one side in order to give a clear view of vessels approaching from astern which have right of way.

As recently as the early years of this century the bridge was often still very basic, and a captain could usually be seen leaning over the bridge wing as a vessel docked; on any sizeable vessel today officers are as likely to be inside scanning closed-circuit TV monitors while communicating directly with the dock-ing crews fore and aft by shortwave radio and operating the ship's side-mounted bow thrusters directly. The ferries plying between Stockholm and Helsinki have set new standards, not least in such innovations as the fully-integrated navigation system fitted to the 1990 *Silja Serenade* (seen overleaf with her sister ship *Silja Symphony*), largely devised by her Captain. The overall design of bridge fronts on these vessels illustrates a balance between navigational priorities in demand-

*OPPOSITE, FROM ABOVE, L to R: A floating container for self-propelled cargo – the car carrier Nosac Sea; Function with style – the Irish bulk carrier Arklow Brook from Appledore Shipbuilders on trials, December 1995; Fully equipped for cargo handling, but with containers already on deck, the 1963 cargoliner Glenfalloch. She and her three sisters are often regarded as the finest cargo vessels ever built; The 1973 tanker Globtik Tokyo makes waves in ballast. The figure on the fo'c's'le has a better view than the captain on the bridge of the aircraft who took this picture; Directionality in unlikely circumstances – a small modern vessel at Teignmouth docks, 1995; The labour-intensive detail of older vessels invites detailed examination – the deck and bridges of the 1929 Orontes; FROM ABOVE: Fully-enclosed lifeboat on a Russian bulk carrier, Tilbury; Clean modernity – stern of the 1994 new generation cable vessel Asean Restorer from Kvaerner Masa-Yards; Harmony of form and function – the clipper bow of this elderly cable-laying vessel ensures that the cable is kept clear of the hull, Plymouth docks, 1964*

ing waters and the commercial priority of offering passengers space, as well as views of the spectacular scenery.[33] On some cruise liners the bridge no longer dominates the front of the superstructure but is surmounted by an observation deck, or lounge, for passengers.

On many modern passenger vessels the bridge itself is streamlined into the superstructure front and so loses much of its traditional prominence as a command centre. An exception here is P&O's new cruise liner *Oriana*, which retains traditional open wings to her bridge, located directly below the passenger observation deck, but only because these were specified by P&O's Chairman, Sir Jeffrey Sterling, so that passengers could watch the ship's officers controlling aspects of docking manouevres in port. As so often is the case, functionality in ship design is complex.[34] Basic understanding of ships' functional design shapes our response to their appearance, and to some extent it does with all architecture which does not deliberately conceal its function. Our response to a cathedral is likely to be diminished (or at least inappropriate) if we think the building a hotel, or are simply ignorant as to its use.

The proportional relationship between the main elements of ship design – hull, superstructure and gear – is determined more by considerations already outlined than by aesthetics, but like a skilled mathematician a good designer will find a solution not merely correct but elegant. Solutions will differ from ship to ship, and in distinguishing between the design of passenger cruise vessels and of cargo vessels, Gianfranco Bertaglia compares the latter with surgical instruments created specifically to do a very particular job well, accepting that an efficient tool has a beauty all its own. The British marine artist Peter Anson responded in just this spirit to the arrival of the *Duchessa d'Aosta* in Venice in the summer of 1926:

> I strolled down to the Piazza San Marco and beheld . . . a 7000-ton cargo vessel of the Navigazione Libera Triestina just coming round the angle of the Palazzo Ducale, her name painted amidships in block letters. I liked the look of her at first sight. Her four two-legged lattice masts gave her definite character, for it was obvious that she had been designed for functional purposes and not for effect. For this reason she possessed real beauty of her own.[35]

With all passenger vessels, however, special consideration will be given to exterior and interior appearance since prospective passengers exercise choice as to which ship they travel on. Scheduled passenger ferries on a short crossing would perhaps be at the bottom of such a scale (although increasingly they present a very self-conscious profile); today's luxury cruise liners are at the top, needing as they do a style which proclaims luxury and modernity.

A ship's hull functions, and is perceived as, the main mass of a ship to be propelled through the water. It has in essence all the simplicity and logic of a heavy log pushed across the waters of a still pond: rounded at the bow, weighted with cargo, and with engine, propeller and rudder at the stern, its simplicity underpins all our more complex response to a ship in motion, especially the pitching and rolling of a vessel in heavy weather. In fast naval craft the visual impact of a fine-lined hull designed for speed is enhanced by the contrast of block-like superstructure, and of extensive communications and weaponry which can make no concessions to aesthetics. Recently, however, some warship super-

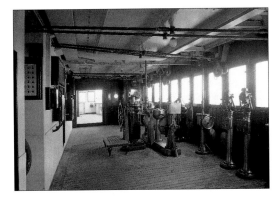

*OPPOSITE, FROM ABOVE, L to R: The 1935 Dutch coaster Venus has the straight stern and counter stern of earlier steamers, although she was built with diesel power, Brixham, August 1959; The typical Dutch coaster Bram was built in 1956, again with diesel, but has rounder lines, Brixham, August 1959; The 1976 Jugoslavian tanker Jadran in ballast; The telescopic bridge structure (here lowered) of the German low air-draught cargo vessel Osterberg; The Osterberg leaves port with bridge raised, Teignmouth, 1995; In command – the wheelhouse and bridge of the Nelson Line Highland Loch, built in 1911. The vessel ran from Britain to the River Plate and the wooden beams in the foreground carried sun awnings when needed; FROM ABOVE: State of the art – the sparse and spacious bridge of the 1913 White Star liner Aquitania; The paddle steamer Monarch, 1924, pictured in Poole harbour, 1959 (with classic Ford Zephyr), Cosens and Company Ltd ran excursions along the south coast of England; The sister ships Silja Serenade and Silja Symphony demonstrate the visual impact of different superstructure styling and window layout, Helsinki, 1995*

*OPPOSITE, FROM ABOVE, L to R: Boldly-shaped panoramic passenger space tops the superstructure of the* Costa Romantica; *Traditional bridge wings and wood trim are a feature of P&O's 1995 liner* Oriana, *but technology is also evident; Traditional lifeboat and davits on the 1943 standard wartime coastal tanker* Audacity, *Brixham, 1960; The very basic wheelhouse and canvas-edged bridge of the steam dredger* Plym, *Plymouth Docks, 1963; FROM ABOVE: Allan Sekula, 'Panorama, mid-Atlantic' from* Fish Story, *1995; The decisively-functional form of aircraft carriers impresses by scale and contrast – this is the USS George Washington at Spithead*

structure has assumed a more angled profile in order to minimise detection by radar, an unusual example of invisible processes shaping the visible profile of a vessel.

A major difference between ship and land architecture is that in order to function ships have to be designed fully from all sides (a following sea can be just as dangerous as one into which the vessel heads); they also present a dramatically different appearance from various viewpoints. In its active elements a ship is, in terms of the distinction that *Learning from Las Vegas* drew between 'duck' and 'decorated shed', appropriately a duck: a form where 'the architectural systems of space, structure, and program are submerged and distorted by an overall symbolic form'.[36] All ships change dramatically when viewed from the side (amidships), from the front (ahead) and from the rear (astern). Conversely, some buildings – both old and new – offer a relatively similar appearance from any perspective. Only ships such as the double-ended ferries which ply the harbour routes of Sydney and Hong Kong, or shortsea routes such as that between Helsingfors and Helsingborg, do not present rapidly-changing shapes to the observer.

A rare imaginative approach to ships and the sea is Nöel Mostert's *Supership* (1974), a study poised eruditely between informed and affectionate appreciation of maritime heritage and a wondering – and worried – assessment of the large-scale and technology-based shipping world represented by the P&O tanker on which he travelled, and around whose voyage his book is based: 'These huge, strange, unattractive ships struck me as a new and significant part of man's story on the waters'.[37] In its breadth of interest, and above all in Mostert's use of maritime history and detail for the wider and significant insight, *Supership* splendidly illustrates Alain Corbin's later insight in *The Lure of the Sea* (1988) that 'the history of seafarers and navigation, however prestigious it might be, is not the best way to understand and analyse images of the sea and its shores'.[38] Corbin's view, unsettling as it may be for some maritime historians, is also splendidly confirmed by Allan Sekula's *Fish Story*, which brings contemporary photography and cultural analysis to bear both on global capitalism and intrinsic images of seafaring trade from Dutch seventeenth-century painting to today's supertankers:

What one sees in a harbor is the concrete movement of goods. This movement can be explained in its totality only through recourse to abstraction. Marx tells us this, even if no one is listening anymore. If the stock market is the site in which the abstract character of money rules, the harbor is the site in which material goods appear in bulk, in the very flux of exchange.[39]

Sekula's photographs both record and probe that materiality, and nowhere more dramatically than in his mid-Atlantic picture from the bridge of the container ship *Sea-Land Quality*.

The story of modern shipping is very much of technology, visible and invisible, going to sea and the consequent disaffection of many interested in ships at both an amateur and a scholarly level has been both unfortunate and misguided; it has discouraged that understanding of ships as mobile architectural structures which this study seeks to advance. This is the more ironic since Isambard Kingdom Brunel, a key figure in the development of the modern ship, was interested in the engineering of both land and sea transport; in that, too, his plans foreshadowed today's global transport network.[40] Brunel was himself much influenced by that

*FROM ABOVE: The* Alabarda *is a Fincantieri study for a top-class cruise ship of decisively modern design; Ship basics – buoyancy, carrying capacity and controlled propulsion – dugout canoe, Bengal, 1963; Functional form is different in and out of the water – a fast gunboat from Kvaerner Masa-Yards, Finland*

most innovative and practical building, Joseph Paxton's cast-iron structure for London's Great Exhibition of Art and Industry of 1851, known popularly then and since as the Crystal Palace: 'no talk of architecture again can ever exclude it'.[41] Yet the Great Exhibition title, 'Art and Industry', reflected a class-related divide in British culture which Brunel himself felt and which still causes hostile distinctions between supposedly 'mechanical' and 'artistic' studies and professions, as well as between art history, architecture and design. In continental Europe these disciplines work more happily together in many fields, not least ship design and shipbuilding. My term 'ship architecture' deliberately encapsulates the tensions in Britain, because from the first such tensions also affected attitudes to photography, an 'industrial' or 'mechanical' medium as central to the aims of this book as to its production.

Paxton's building is today the most interesting artifact of the 1851 Exhibition, literally containing all other exhibits within its overarching transparency yet itself neither 'art' nor 'industry', but defying classification. Although destroyed by fire in November 1936, long after its dismantling, re-erection (and extension) at Sydenham in 1852-54, its spirit lives on in the spans of Brunel's 1854 Paddington station in London, and also in key modern buildings.[42]

As an outside winner of the competition for the Great Exhibition building Paxton's design attracted considerable hostility, not least because he himself was not an architect – nor even an engineer. He had designed large and successful glasshouses for private clients but such structures were regarded as 'culturally invisible'; one contemporary critic dismissed them as 'no more subject to the rules of civil architecture than is a ship'.[43] As John McKean notes in his 1994 study of Crystal Palace, this is precisely why the Crystal Palace 'can stand unembarrassedly unadorned, as a naive but precision instrument pretending to be free of emblematic power'.[44] Ruskin's view of architecture as decorated construction clearly excluded the Crystal Palace from consideration, but:

> there are other real arguments: Crystal Palace is not architecture because it is not permanent. On the other hand, engineering can move, is glimpsed, passes at speed. Architecture is located, it enriches a *genius loci*. Crystal Palace, like its tiny, distant descendant the Sainsbury Centre at East Anglia University, just lands on any field.[45]

Such considerations bear upon our understanding of modern ship architecture in a wider context. Totally mobile and always functional, ships are also transient structures, notwithstanding their complexity and scale; built to move and to work efficiently, if they survive natural disasters they are scrapped and recycled or, occasionally, rebuilt. The liner *Italia Prima* has been reconstructed on the hull of Swedish Lloyd's 1948 *Stockholm* in Genova, home port of Italy's *Andrea Doria*, which she herself sank in a collision off New York in 1956.[46]

Rarely is ship architecture associated with the name of a designer and when it is – as P&O's 1990 cruise liner *Crown Princess* is with that of Italian architect Renzo Piano – the name does not always tell the whole story. Piano can more accurately be described as having 'styled' a ship which was designed (to be safe and seaworthy, amongst other things) by naval architects at the builders, Fincantieri.[47] Whether or not Sir Norman Foster's Sainsbury Centre can be seen as having landed on an East Anglian field, perhaps from his new Stansted Airport,

*FROM ABOVE: Aground on the causeway at Tilbury, on the Thames, the 1891 steam coaster TG Hutton shows her straight stem, counter stern and open bridge. The elegance of such vessels lies in their simplicity of outline and the proportions of key elements. She was refloated safely; The steam coaster Yewcroft was built in 1929, some forty years after the TG Hutton, but shows little development. She ran aground off Cudden Point, Cornwall, on 8 August 1956 and is seen here eleven days later, her back broken; The Russian cargo vessel Demetrius wrecked at Prawle Point, Cornwall, December 1992*

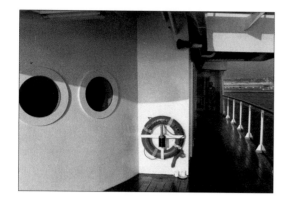

*FROM ABOVE: Pleasing shapes and spaces on a 1994 Scandlines Helsingborg-Helsingor ferry, 1995; Port architecture – two modern Swedish icebreakers with the 1985 passenger and vehicle ferry Mariella, Stockholm, 1995; The tanker Alice PG from Appledore Shipbuilders on trials off the North Devon coast, September, 1994. Such modern vessels appeal through clarity of line, proportion and colour*

mobility is a defining element in ship architecture since it affects not only a vessel's structure and size, but also our responses to it as built object – responses which remain complex whether a ship is under construction, at work, being scrapped, or, most evocative of all, perhaps, if it is wrecked.[48] Unlike most land architecture, ship architecture is traditionally a story not of straight lines but of curves, of varied materials – wood, iron, steel – fashioned by intensive and skilled labour into shapes that both fit themselves to the sea and are visually pleasing. And as Foster reminds us, architecture is only good if it uplifts the spirit as well as being more narrowly efficient; there are lessons here for devotees of narrow 'efficiency' in other areas, too.

Some architects today voice concern at their students' acceptance of increasingly second-hand knowledge of architectural sites, and at a growing emphasis on theory rather than practice. Today tried and proved design systems allow the possibility of modifying a shape to give any desired style or elegance, but overall ships still exemplify an uncompromising materiality, not least since they function by the dynamic of solid against liquid. A building which is overweight is wasteful to construct, but still functions; in a ship surplus steel means a daily penalty in operating costs, and most ships in fact weigh only one third of a building of comparable size. They are the most efficient by weight of large structures but their high premium on efficiency dates them rapidly in the face of technological and economic change, or even of world politics. The 1956 closure of the Suez Canal necessitated the rapid building of tankers large enough to make the much longer voyage around the Cape of Good Hope economically feasible.

On a calm day a well-designed ship looks splendidly at home on the sea, but the Beaufort Scale of wind force is indifferent to aesthetics and ships are built first and foremost to survive: a ship's bow flare allows some scope for aesthetic treatment but its primary function is to pierce the sea and in doing so throw water clear of decks and cargo, while also providing increased buoyancy the deeper the bow sinks.[49] The naval architect Marshall Meek stresses the need always to remember this essentially dynamic aspect of a ship's life:

I've often thought that many shipyards, and many naval architects even, forget just what a ship has to do because they see it fitting out for long periods in a yard when it's static, whereas at sea a ship is never static.

Each element in a ship's active design, namely those elements below the waterline which directly affect its seagoing characteristics, is defined largely by previous experience of unchanging operational conditions; significant changes, and there have been many, will be exhaustively tested at model scale before introduction and then the prototype itself carefully monitored. Sometimes, as with the introduction of the first British container ships, it was necessary to make significant departures from knowledge accrued to date:

When work began early in 1966 on the design of these ships there was not a great deal of precedent to work on. Although container ships were operating in the United States these were mainly conversions of existing conventional tonnage. For the first time designers, accustomed to cargo liner work, found themselves without a basis ship.[50]

A ship is essentially a live and always-shifting structure, but the slow and deliberate manoeuvring of a ship in port reflects its power and mass, a buoyant mass of great inertia which – should power be lost at sea – is also immensely vulnerable: a certain stubborn and pessimistic insistence on the primacy of material forces is part of the common culture of harbor residents. This crude materialism is underwritten by disaster. Ships explode, leak, sink, collide. Accidents happen every day. Gravity is recognised as a force.[51]

Ships have always been familiar machines to the few and remote objects of fear or fantasy to the many: most of us, after all, live on land. One consequence is that the imaginative significance of our culture's largest mobile structures, carriers alike of national pride, regional exports and the individual traveller, vessels which have both served and symbolised complex aspirations, is yet to be explored. The importance of modern ships even in the fields of engineering, architecture and design is commonly neglected, as indeed it is in Academy's own studies, *Great Engineers* (1987) and *Architecture of Transportation* (1994), although Robert Kronenberg's recent study, *Houses in Motion* (1995), does recognise ships as important 'non architectural' precedents for a questioning of the traditional division between 'building' and 'architecture':

> Portable buildings . . . have a unique place in the human understanding of the nature of object and place as they can be artifacts that cross cultural and symbolic boundaries. It is normal for even the most land-bound individual to admire the grace and beauty of a sailing ship; its purpose is evident in its shape and form.[52]

Kronenberg illustrates early twentieth-century liners, modern aircraft carriers and some specialist vessels in his study, but reference to 'the grace and beauty of a sailing ship' reinforces a generally nostalgic attitude to ship aesthetics. If ships have a role in contemporary architectural debate then grace and beauty should be as apparent as much in modern as in past examples; taking sailing ships as the standard tends to perpetuate that very tendency to see pre-industrial vessels as beautiful, later vessels as 'merely' functional, which feeds the art versus industry divide. Both considerations must inform the design and function of ships and our understanding of it. Kronenberg himself touches on this issue elsewhere in quoting Nicholas Goldsmith on the problems of constructing a 1991 mobile auditorium in New York:

> Our design approach to this project was to let the engineering of the mechanism dictate the forms and geometry of the structure . . . The architectural poetry was in the proportions and the relations of these elements to each other. The design became a mixture of architecture, industrial design and engineering . . . This project put us on the edge of our profession by raising the question: When is a structure a machine and when is it a building?[53]

That writing on ships has generally failed to consider how, in a contested post-colonial and post-modern era, the design and function of ships might relate to broader cultural patterns is regrettable, the more so as passenger ship designers and operators are themselves alert to these issues for good commercial reasons.

Here Allan Sekula's synoptic vision of ships from a sharp visual and political perspective is invaluable, as is that of Jonathan Raban, who had never sailed before publishing his incisive and entertaining book *Coasting* (1986) in the turbulent cultural wake of the Falklands War. Raban has elsewhere observed of his own

*FROM ABOVE: Tank-testing a ship model at Kvaerner Masa-Yards, Finland; Architecture that moves is always at risk – the damaged bow of the cargo vessel* Potaro, *Southampton late 1950s*

country's maritime writing that in the nineteenth century it 'was aggressively reactionary and backward looking', that the political colour of late Romantic writing about the sea – including such salty heroes as Conrad and Masefield – was 'a very deep blue',[54] and such factors still influence maritime research, collecting and display policy, especially as revenue generation assumes increasing importance. Indeed, Sekula argues that there are real problems with the very nature of maritime subjects:

> Most sea stories are allegories of authority. In this sense alone politics is never far away. The ship is one of the last unequivocal bastions of absolutism, regardless of the political system behind the flag that flies from the stern, or behind the flag or corporate logo behind the flag that flies from the stern.[55]

Avoiding the easy nostalgia of much writing on ships, this book seeks to elucidate the design of some modern vessels, not only for the specialist but also for the general reader seeking an informed appreciation of the world we make for ourselves, and of which ships are a vital part. Most writers on design favour Italian cars and coffee machines, yet nearly all world trade still moves by ship – many of them also built in Italy – including containerloads of those very computers which enable some to inhabit a seemingly timeless world of cyberspace. Such self-deception Sekula terms 'forgetting the sea', arguing for 'the continued importance of maritime space' in countering 'the exaggerated importance attached to that largely metaphysical construct, "cyberspace", and to the corollary myth of "instantaneous" contact between distant spaces':

> I am often struck by the ignorance of intellectuals in this respect: the self-congratulating conceptual aggrandizement of 'information' frequently is accompanied by peculiar erroneous beliefs: among these is the widely-held quasi-anthropomorphic notion that most of the world's cargo travels as people do, by air.[56]

Against what might be termed such intellectualisation of world transport, Sekula contends that 'Large-scale material flows remain intractable' and that:

> the proliferation of air-courier companies and mail-order catalogues serving the professional, domestic, and leisure needs of the managerial and intellectual classes does nothing to bring consciousness down to earth, or to turn it in the direction of the sea, the forgotten space.[57]

Any study of ships as an aspect of twentieth-century architecture must address the unique significance of the ship in Western art and society. At least since Noah's maritime ventures in livestock transport ships have been symbols of salvation, but admiration for them has always been tempered by awareness of the daily risks posed by seafaring itself; in this respect the fortunes of a ship and her crew mirror our own more closely than do those of any other constructed object. No land-based structure, however impressive and howsoever destroyed, can even approach our continuing fascination with the *Titanic* slipping – apparently intact and with lights ablaze – beneath the dark surface of the North Atlantic on 15 April 1912. That single event encapsulates simply and horrifically the sea's effortless ability to swallow constructions of great sophistication whose craftsmanship expresses the confidence of those that build and sail them that what, in fact, happened to the *Titanic* will not happen. The fact that so much of the *Titanic* survives on the seabed greatens rather than lessens the shock, as Thomas

Hardy foresaw in his poem, *The Convergence of the Twain*, in which seaworms travel over unbroken gilt mirrors in which the rich and famous should have admired themselves. Some awareness of these conflicting appeals, no matter how vague or unspoken, informs the process of designing, travelling on, or simply watching a ship. As Mostert expresses it:

> The sea's horizon was the awful edge of the proven universe, and of the unproven beyond. The ship – so perishable upon those infinite and unknown waters – was the true microcosm, and sailors the most fateful of men, confronting from their decks, unprotected by the myriad shades, niches, and succours of the land, the fullest view of the whims, confusions, perversities, and sheer helplessness of existence.[58]

This passage captures well the post-Romantic response to the sea and ships, but there are other questions, and especially those relating to the corporate investments needed (then as now) to build and operate ships, which such evocative writing does not touch and which challenge many assumptions of maritime heritage and its attendant political values. As Friedrich Engels wrote in 1844 of London, approached via the River Thames and its 'concentration of ships':

> The traveler has good reason to marvel at England's greatness even before he steps on English soil. It is only later that the traveler appreciates the human suffering that has made this possible.[59]

Recognition of this human dimension of ship architecture – in both its construction and operation – prevents any slide into nostalgia: 'Culturally, the sea has become a vast reservoir of anachronisms, its representation redundant and overcoded'.[60] Maritime heritage as such lies beyond the scope of this book, but cherished national traditions profoundly shape attitudes to ship architecture, from Trafalgar to The Falklands. More broadly, Gianfranco Bertaglia at the Italian shipbuilders Fincantieri sees ship design as 'a process of educated imagination, a type of culture that can exist and survive well in the environment of a large industry with long-established traditions'.[61] In contrast, Marshall Meek commented as follows on Fincantieri's design for P&O's 1997 superliner *Grand Princess*, in which Bertaglia exploits diverse historical and contemporary references:

> I can more readily appreciate an Italian talking in that kind of way; shipyards here talk basics. That's just typical of – I wouldn't say what's gone wrong – but the nature of the shipbuilding industry in this country. They haven't been ready to adapt at all. We lose something here compared with the Italians, who are prepared to look at things in a different way.

Le Corbusier once remarked that 'We have no right to waste our strength on worn-out tackle, we must scrap and re-equip',[62] and, when not content to run 'worn-out tackle' for a handsome profit, this has been the attitude of the shipping industry itself. Physical deterioration, maintenance costs and sheer obsolescence usually make it preferable to sell, or to scrap for new tonnage, rather than modernise elderly vessels. The financial calculations involved today in a replacement cycle can be daunting; a new container vessel with its three sets of containers will cost around US $100 million and may have a life of some thirty years in prime condition, not much different from the vessels of the 1890s, given the economic pressures of modern shipping. Today's vessels, however, are far more efficient both in carrying capacity and in propulsion, just as the 1890s steam vessel itself

represented a considerable advance in efficiency over earlier sailing vessels.[63] Container ships may last longer than many other vessels because their cargo is fixed and clean, and load stresses can now be accurately calculated.[64]

A ship lives and moves in an environment calculated to destroy it. Storms, rocks and human error aside, salt water and electrolysis do to steel hulls what the marine borer worm does to wooden ones. The loading and unloading of cargo, as well as the relentless bump and scrape of docking manoeuvres, abrade protective paintwork, while the strains imposed on hull structures by waves, engine vibration and ever-changing loads and stress are intense. Some modern ocean-going ships have broken apart and sunk, and whole bow sections fallen off, while other vessels have needed extensive repair and strengthening. The need for vigilance and maintenance is constant and size brings its own problems, not least complacency. Raban was duly unnerved by an officer's comments as his container vessel *Atlantic Conveyor* butted its way from Liverpool to Halifax in 1988:

> The wind was blowing from west-northwest at Gale Force 8 and gusting to Severe Gale 9. Vince, the officer-of-the-watch, put the swell at 30 feet, though from the bridge it looked less. 'That's the danger of a ship like this. Up here, you're so far removed from the elements, you get blasé . . . and in this sort of weather you have to know exactly what you're asking a ship to do. Otherwise you'll overstretch her.' I didn't care for the sound of 'overstretch'. Studied through binoculars, the torque looked painful.[65]

Land architects have more often seen ships in terms of their own preoccupations than sought to learn from maritime demands and constraints. The ship references routinely cited in architectural and design history prove delusory: those of the Italian Futurists stress superficial style with quite impractical size and speed while showing no understanding of, nor interest in, the practical complexities of an ocean-going ship. Similarly, Le Corbusier's section, 'Eyes Which Do Not See', in *Vers une architecture* (1923) draws attention to the importance of 'good proportions' and 'the play of . . . masses and the materials used' in contemporary ocean liner design. His illustrations are of Cunard's 1913 liner *Aquitania*, but his reference to 'these formidable affairs that steamships are' is naively uninformative. He praises a lounge on the *Aquitania* for its 'wall all windows, a saloon full of light', but is uninterested in how such a passenger space – hardly typical of all ocean-going vessels at the time, or even of passenger saloons for all classes – might be functionally located, given the demands of designing and building such a vessel. The *Aquitania* was by no means all windows; everything depended on where you were, and that depended on which class you were in (at least on board). By contrast, the British architect James Gardner (who had never designed a ship before) recalled that the main consideration for him in styling the above-water form of the *QE2* in the late 1960s was the North Atlantic, even though the vessel was intended to spend its winters cruising in kinder waters. In the summer of 1995 the liner encountered a 95-foot Atlantic wave and survived undamaged; the main credit for this must go to good empirically-derived engineering design in the offices of Upper Clyde Shipbuilders and of Cunard, hull tank testing, and meeting the requirements of Lloyd's Register of Shipping. That Gardner's own attitude as a designer was attentive to such practical issues is impressive. By contrast, Le Corbusier's claim that the steamship was 'the first stage in the realisation of a world organized according to the new spirit' ignored the specific and volatile mix of commerce and politics which shaped and maintained such vessels. It is a sobering judgement on his theories that developments in the shape of the jet airliner (as they were once termed), would render such passenger liners (for transport as opposed to cruising) obsolete within fifty years.

Not surprisingly, perhaps, much writing on ship design tells us more of its author and period than of ships themselves. When Marinetti writes in the 1909 Futurist *Manifesto* of 'great steamers that scent the horizon' he reflects his own essentially urban fascination with ocean voyaging, his own imaginative remoteness as a land-locked visionary indeed condemned to 'scent the horizon'.[66] Interestingly, Le Corbusier's references to ship design urge us to 'forget for a moment that a steamship is a machine for transport and look at it with a fresh eye'. In this way, he claims, a 'seriously-minded architect, looking at it as an architect, (ie: a creator of organisms), will find in a steamship his freedom from an age-long but contemptible enslavement to the past'.[67] The ship is here invoked as a liberating image, but only if regarded by an eye that is 'fresh', namely devoid of any interest in how such a vessel is built, or operates as a mode of transport. As Sekula suggests, ships can 'function both as prisons and as engines of flight and escape',[68] and much turns on the aspirations of the writer.

If commentators can be unreliable, the attitudes of mariners themselves are not always what those on shore might wish. Joseph Conrad, a captain on square-riggers in the China Seas before he was a writer, felt (at least in the character of his narrator, Marlow) that most sailors 'lead, if one may so express it, a sedentary life. Their minds are of the stay-at-home order, and their home is always with them'. It is true in one sense that 'One ship is very much like another, and the sea is always the same'.[69] To design and build a ship is, consciously or not, to contribute to a longstanding tradition of human artifice and, unlike other constructions of comparable size and diversity, all ships invite comparison with one another through their shared requirements of buoyancy and mobility; these powerful associations explain the lasting fascination they exert over the human spirit. To understand the iconography and design of ships, however, is also an exercise in what Bertaglia terms 'educated imagination'. Such education of the imagination requires 'a culture both broad and profound, with excellent recall of past experience and a sure grasp of the present', and it is exactly this informing sense of 'education' with regard to ships that is lacking in the writings of the Futurists and of Le Corbusier.

Ships are built, operated and scrapped out of sight. Time in port is kept to a minimum: shipowners 'prefer to have their ships at sea, where they cost less to run . . . as soon as a ship enters port a great deal of energy and initiative is used to get her out again'.[70] Much investment and energy is devoted to positioning the ship functionally as 'nothing more than a squat and motionless floating warehouse',[71] and the increasing brevity of such visits means that vessels today form little permanent part of that 'visible cultural landscape' which John Heskett sees as the province of design debate.[72] In Heskett's own terms, though, they do feature significantly in the 'complex pattern of function and meaning' which shapes 'our perceptions of the world', and not only visibly. As Raban notes, the terminology of ships and the sea pervades every aspect of our language:

Modern English is littered with dead nautical metaphors which were alive

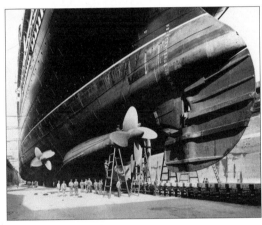

*FROM ABOVE: Wooden scaffolding around the* Aquitania *on the stocks at John Brown's, Clydebank, April 1913; The* Aquitania *in Gladstone Dock, Liverpool, May 1914, prior to her maiden voyage. She was launched on 21 April.*

*FROM ABOVE: An uneasy juxtaposition of rattan and rivets – the* Aquitania's First Class Promenade, A Deck; *Rather basic facilities on the* Aquitania's Second Class Promenade, A Deck; *The flare of the hull itself partly defines the volume of the* Aquitania's Third Class Promenade, D Deck

and well when Shakespeare was writing. To have things 'above board' . . . to be 'taken aback' . . . to see something out 'to the bitter end' . . . the most landlubberly speaker of English is prone to talk unconsciously in terms that come out of the sea.[73]

For many peoples the ship also retains its historical power as an image of salvation, as witness the 1992 photograph by Oliviero Toscani of Albanian refugees covering every surface of a cargo ship arriving in the Italian port of Bari, an image used as a poster by Benetton. Sekula writes of other such fraught voyages of our times:

Consider . . . the glacial caution with which contraband human cargo moves. Chinese immigrant-smuggling ships can take longer than seventeenth-century sailing vessels to reach their destinations, spending over a year in miserable and meandering transit. At their lowest depths, capitalist labour markets exhibit a miserly patience.[74]

The practicality of most merchant ships; their location at sea; their perception as products of industrial grime and of shared (anonymous) engineering 'skill' rather than high-tech (individual) architectural 'design' – above all, a lifespan which means that a 'grand tour' even of 1960s vessels is problematic – all these factors locate ships firmly outside 'architecture' in both public and professional debate. Few people could name a ship designer apart from Brunel, and he proudly considered himself first and foremost an engineer. Ships today are often either incomprehensible or remote; boring nuts-and-bolts industrial products or fantasy objects inaccessible by virtue of location or expense. Yet despite, or perhaps because of, such experiential distance ships retain a special cultural reference, not least because of that hubristic defiance in transience of which I have suggested the *Titanic* is the classic example: 'Ships, after all were, are, the buildings of the water, and they grew in much the same way and with much the same care and faddishness as the buildings of the land'.[75]

If the economic function of merchant ships remains greater than generally recognised, in the still-growing cruise market ships increasingly offer special and specific versions of fantasy. Even modern ferries have claimed this role as far as their size and routes allow, and 'fantasy' and suchlike terms now feature prominently in their names. Fantasy is, of course, a significant part of the highly profitable worldwide leisure industry, and those aspects of modern cruise ships which provide it are supreme illustrations of the impossibility of any simple contrast between 'efficiency' and 'decoration'. A vessel whose decor attracts and pleases fare-paying passengers is undeniably more efficient in cash terms than one whose supposedly stark and 'functional' design – no matter how acclaimed professionally – does not.

We have seen that the closest parallels here in land architecture are perhaps in the desert city of Las Vegas, and Alan Hess's reading of Vegas style in the 1940s provides a useful approach also to liner interior design, both then and later:

This was theme architecture: the thorough depiction of a particular historical era or geographical area in the architecture, ornament, costumes, and service of the hotel . . . The re-creation does not have to be accurate; it usually tells us more about the era that produced it than the era that it purported to depict. Often nostalgic, it is a way for the fast-moving present to achieve a level of comfort by surrounding itself with a known past.[76]

Shopping malls have been envisaged as 'plunked down in the middle of an enormous asphalt sea', as 'pedestrian islands in an asphalt sea',[77] but Las Vegas has been described as a whole city which, architecturally at least, has let go of its moorings and it seems fitting that over the years several buildings have been in the shape of floating gambling houses – Mississippi riverboats. More recently high-tech fights between pirate ships have been the centre of displays at one Las Vegas hotel.[78] Yet the high-tech and custom-designed fantasies offered by the secluded interiors of some modern cruise liners also draw upon a long tradition of Western voyaging in search of the exotic:

> The sea has always exerted a strong fascination for the intuitive tradition because it is the last place on earth still undomesticated by man and still relatively dangerous.[79]

Another United States city which, logically enough, invites comparison with the design and function of today's cruise vessels is Miami, 'home port' for many of the world's most modern and luxurious ships. It is odd to think of many of those for Carnival Cruises being launched in wintry Finnish ice, only to spend their working lives in a sunshine context whose architectural traditions in many ways parallel their own:

> The entire city is meant to be a quote from the past; it integrates all the references, all the models that come together to forge its true identity . . .in Miami, there is a coexistence of architectural values from the 1920s . . . the 1950s . . . the Atlantis building of the 1980s by the Arquitectonica; the Haitian market of 1990 . . .[80]

It is perhaps bizarrely appropriate to the age of cyberspace and space travel that, whereas in earlier periods one might take a ship to exotic ports, today the modern cruise liner itself contains scenic multitudes; passengers can Caribbean-island-hop in surroundings which include Egyptian-style décor realised with the advice of Egyptologists to ensure accuracy and five kilometres of computer-controlled flexible fluorescent lights to create an electronic dawn in the vessel's multi-storey atrium. Such sophistication promises the passenger almost total isolation from any world that might pass as reality in order to make discoveries on board the vessel itself, not ashore. Joseph Farcus lives in Miami, understands this principle well, and applies it successfully in his cruise ship designs for Carnival Cruises:

> eclectic individual experiences have also been achieved within this unifying framework, which I feel is the most paramount part of cruise ship design. This encourages and allows the guests to have an awareness of their architectural environment, and enjoy a discovery process throughout the cruise.[81]

*FROM ABOVE: The ship as salvation – Albanian refugees cover a ship arriving at the Italian port of Bari, an image used as a Benetton poster, 1992; The ship as nuts and bolts – assembling the steamship* Robert Corydon *prior to its dismantling and shipping to Africa, where it was to be reassembled for inland lake service; The ship as affordable fantasy – the 1993 Carnival Cruise Lines vessel* Sensation

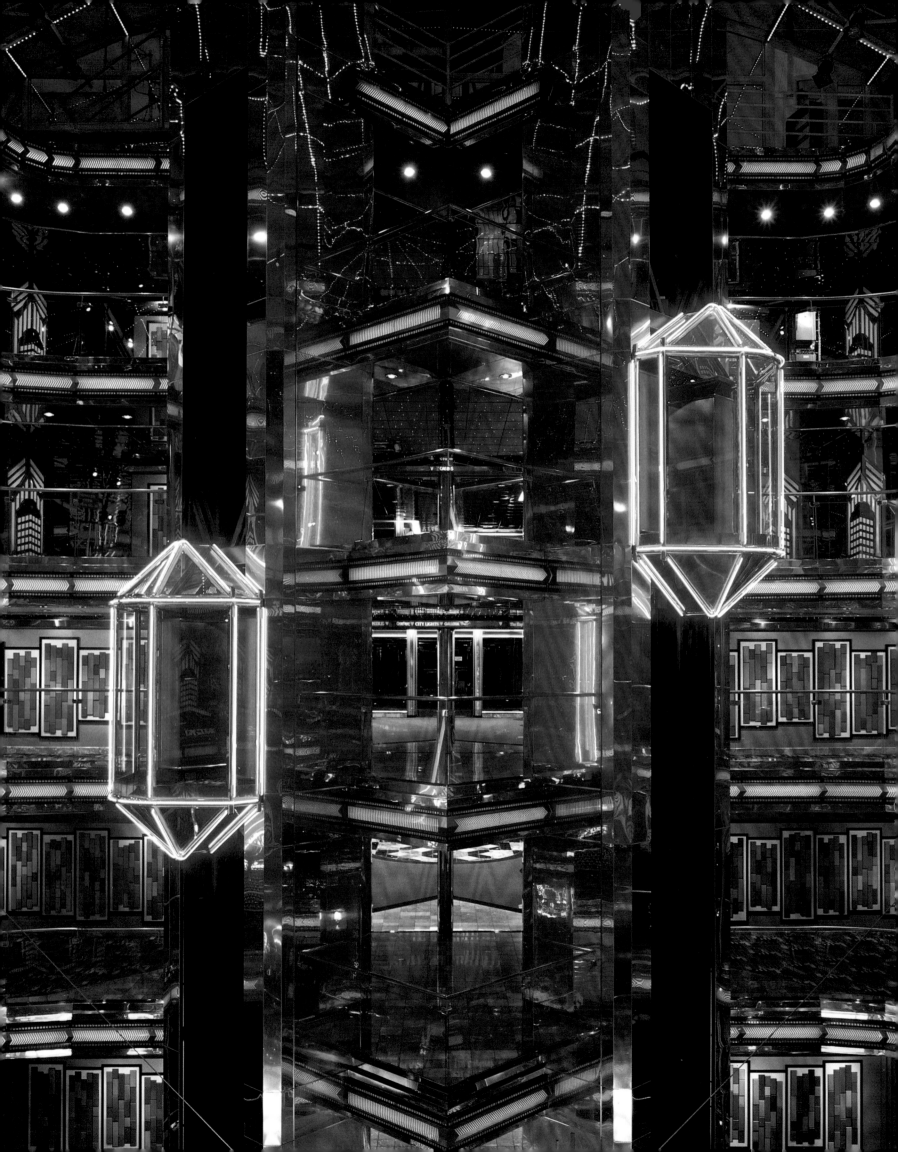

# FORM, MEANING AND THE
# MARITIME POST-MODERN

*The ship is now more likely to be a unit in a functional ensemble encompassing
the container crane, the pipeline, the port itself, the railroad. This functional
embeddedness reduces the charm of ships, blunts the mutinous longing
that draws the eye to their form.*

*Allan Sekula,* Fish Story, *1995*

In land architecture recent debate has involved disagreements of both definition
and application between the advocates of modernism, late modernism, post-
modernism and their various combinations. Issues both conceptual and practical
shape the literature of architectural critical debate, and the main positions are
usefully covered by Margaret A Rose in her 1991 study, *The Post-Modern and the
Post-Industrial*, which notes that the term 'post-industrial' dates from 1914, and
'post-modernism' from 1934; both bring a considerable and contentious history
with them.[1] Ships have traditionally not featured in this debate as either architecture or design, but we might make a basic distinction between the mostly functional design of cargo vessels and the increasingly inventive and promotional
styling of passenger cruise vessels in recent years as companies compete for a
well-heeled, well-travelled and style-conscious sector of public leisure. Cargo
vessels appeal to commercial shippers who look for reliable and competitive
rates of transport rather than style, whereas passenger ships appeal to individual
passengers who pay not for the transport of goods but for their own pleasure –
and as the pleasure of the voyage itself increasingly outweighs any notion of
destination it becomes ever more important to travel well rather than to arrive.
That said, cargo companies take account of their corporate image, in which the
design and general appearance of vessels feature significantly (no company
wants to operate a spectacularly ugly ship), and naval architects have their own
personal standards to satisfy in designing a vessel.

The contemporary Swedish designer Robert Tillberg, whose experience goes
back to work on the elegant Swedish Lloyd liner *Kungsholm* (1964), describes a
ship as 'the greatest compromise in the world', given both the complexity of the
enterprise and the myriad constraints to which ships are subject. That the
*Kungsholm* is still sailing (although much altered externally) as P&O's cruise liner
*Victoria*, is one measure of good basic design. Changes in the passenger market
have greatly changed the trade and Tillberg stresses that modern cruise vessels
have nothing in common with earlier liners which operated essentially as a mode
of transport on fixed routes and to regular timetables. A ship that functions
primarily as a floating base for leisure and recreation, perhaps for several weeks,
must meet very different demands from vessels which, however luxurious, offered a transport service between major ports. Joseph Farcus confirms this:

*OPPOSITE: The Grand Atrium on the appropriately-named
1991* Ecstasy *of Carnival Cruise Lines (compare top image,
p12); FROM ABOVE: In drydock the 1939 liner* Angelina
Lauro *(built as the* Oranje*) shows off a sweeping bow flare
and sheerline no longer found on modern passenger vessels;
A model of the 1968 container vessel* Encounter Bay
*illustrates the contrast between the angularity of the cargo
and the fine lines of the hull – a challenging new aesthetic*

Since aircraft dealt a death blow to ships being a method of transportation in the late 1950s and early 1960s, it has become economically necessary to minimize that function.[2]

Quite apart from the practical design issues raised by such specific shifts, many unquestioned and unquestioning attitudes towards ships and the sea can usefully be seen as central to a broader cultural and political debate. Allan Sekula, however, maintains that although 'many of the features of pre-modern and romantic attitudes towards the sea are no longer credible, they surface nonetheless, as if the sea were indeed a bottomless reservoir of well-preserved anachronisms'.[3]

Much maritime writing and research discusses the 'golden age' of sailing vessels in a way which ignores commercial and political considerations, and I have already suggested that acceptance of such ships as setting a general standard of design is misleading. This 'deep blue' approach, as Jonathan Raban terms it, sees maritime space as 'defined in many of its distinctive features by an earlier pre-industrial capitalism, a capitalism based on primitive accumulation and trade'.[4] The arrival of industry is exemplified by the noisy, sooty, greasy machinery of steam, and Joseph Conrad's description in *The Nigger of the 'Narcissus'* (1892) of his ship being towed out at dawn by a steam paddle tug is quoted as indicative by both Raban and Sekula in their respective explorations of maritime iconographies:

> She resembled an enormous and aquatic black beetle, surprised by the light, overwhelmed by the sunshine, trying to escape with ineffectual effort into the distant gloom of the land. She left a lingering smudge of smoke on the sky, and two vanishing trails of foam on the water. On the place where she had stopped a round black patch of soot remained, undulating on the swell – an unclean mark of the creature's rest.[5]

Contemporary leisure boating, an important contemporary aspect of rural and nostalgic attitudes to supposedly unspoilt environments by a largely urban constituency, draws deeply on the oppositional imagery Conrad deploys here, as do many conservation and heritage initiatives.[6] In important maritime respects Conrad's writings were already nostalgic when they first appeared; Raban notes that 1892, the year *The Nigger of the 'Narcissus'* was published, also 'saw the invention of the diesel engine'.[7] By about 1910, less than ten years after Masefield published his *Salt Water Ballads* in 1902, the turbine had been in general marine use for nearly five years. It was in 1897 that the engineer Charles Parsons had dramatically demonstrated the potential of the turbine by famously speeding his *Turbinia* at an uncatchable 34.5 knots through the lines of that year's Spithead Naval Review; the engine itself dated from ten years earlier.

Writing and research on many aspects of modern shipping since 1900 lacks both the range and the status of research devoted to earlier periods. Sail still occupies disproportionate attention, and where steam is discussed at all it is either naval vessels or transatlantic liners between the wars which fill the picture books. James Steele's 1995 study of the *Queen Mary* is a recent quality example of this genre.[8] A related and more serious problem is that no agreed terminology exists as yet in which to illustrate and discuss developments, constraints and design solutions within the field of ship architecture; ways are still needed of defining and illustrating those elements traditional to ship architecture and of

*FROM ABOVE: The elegant superstructure volumes of the 1949 P&O liner* Himalaya *floodlit at Sydney in 1950; The entirely practical but pleasing forms of the bridge on the small fishing vessel* Theodora, *Helsingborg, 1995*

discussing how they have survived, changed, or become redundant. David Pye's 1950 book was a small classic by a designer and teacher, and both Sir Hugh Casson and Donato Riccesi have also written perceptively as practising architects. There remains, however, a need for developments in modern ship design to be addressed as part of mainstream debate. It is ironic that steam locomotives of the American 'streamline' era continue to feature in modern reference books on industrial design when contemporary passenger and cargo vessels are consistently omitted; the one ship to be included is often Raymond Loewy's restyled *Princess Anne* ferryboat of 1933-36, whose streamlined profile seems better suited functionally to a submarine than a ferry. Loewy also worked on three ships for the Panama Railway Steamship Company in 1936 but typically his work on this project was restricted to interior design.[9]

The form of ships is driven more forcibly by both operating conditions and technological change than that of static land structures. The shift from sail to steam was the most visually dramatic example of such technological change, although, of course, many sailing vessels had their masts removed and were themselves converted to diesel power, as were steam vessels later.[10] The move to motor power also brought significant design change both within the hull and in the layout and function of superstructure. Not only did funnels shrink, and occasionally disappear, but the fact that liquid fuel could be more easily loaded and conveniently stored (a benefit already exploited by bilge storage tanks in Cunard's steamer *Aquitania*), together with less need for ventilation in cooler engine rooms and use of a smaller power unit that could be controlled remotely from the bridge, led to decisive changes in appearance. Within these changes there is nevertheless a continuity of design which is important both to comprehend and to articulate recent initiatives. In this context large passenger vessels, and especially modern cruise liners, offer rich examples which are discussed at length later, but some general principles may usefully be set out here.

Many disputed architectural issues involve the supposed purity of Modernism and contrasting styles of pastiche – styles that the architectural critic Charles Jencks has admired as offering the richly referential 'double-coding' of post-modernism, a coding which requires spectator and user to discern and enjoy echoes (often ironic) of different styles, and at different levels. Most of these issues, as we have seen, apply in ship architecture only to those passenger vessels designed to serve the leisure industry, with whose land-based aspects (hotels, motels, leisure resorts and casinos) they share many functional and design priorities. The practice of ship design in all fields is largely defined by the strongly functional nature of vessels themselves, devoid as they mostly are of external decoration – and always cost conscious in the extreme:

> Beneath anyone's romantic notions about being aboard great ships . . . destined for exotic parts, lies the fundamental reason for building those ships in the first place, quite simply to turn a profit for their owners.

Cruise ships can be appropriately compared with international hotel chains, and with Las Vegas; the sea exercises its own brand of conservatism on some crucial exterior aspects of such vessels but inside their steel shell the imagination of the interior designer is licensed by the very concept of 'casting off' and setting sail. In such a context the most relevant aspects of 'Post-Modernism' are those in Jencks's

*FROM ABOVE: The 1927 General Steam Navigation Company's cargo vessel* Woodcock *photographed off Leith on her maiden voyage from London. Her perfect proportions are set off by a single concession to decoration – a white line. The company was innovative in its early use of cargo cranes; Built in 1940 in Sweden as* Caribe II, *the* Greenfinch *was also owned by the General Steam Navigation Company when photographed in April 1960 entering Felixstowe, then a sleepy rural port. The 'black beetle' tug* Richard Lee Barber *of Yarmouth has the tall smokestack and engine-room ventilators typical of steamers, Lowestoft, 1959*

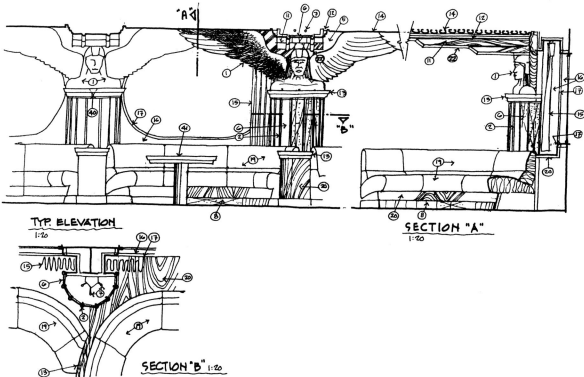

*FROM ABOVE: Atrium on the 1995 Crystal Cruises liner
Crystal Symphony; Drawings by Joseph Farcus PA for details
in the aft lounge on the 1995 Carnival Cruises liner
Imagination*

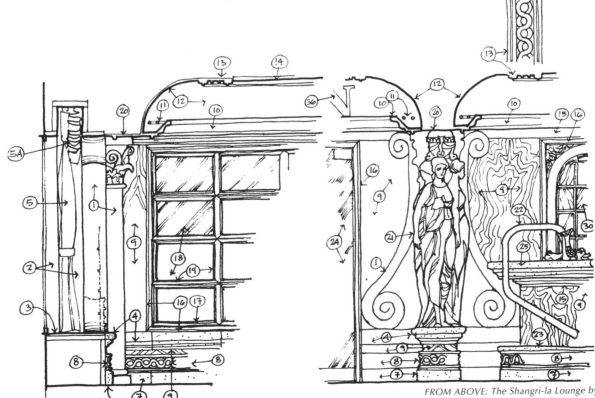

*FROM ABOVE: The Shangri-la Lounge by Joseph Farcus PA
on the 1995 Carnival Cruises liner Imagination; Drawings
by Joseph Farcus PA for decoration in the Grand Bar of the
1996 Carnival Cruises liner Inspiration*

stress on variety of style and reference, along with the benefits of 'a partial return
to tradition and the central role of communicating with the public'.[11] In such a
process, he argues, the most resonant architecture will be one which employs 'a
wide vernacular which includes all sorts of signs and traditional motifs'. Jencks
sees the key issue as only too simple:

> Architects who wanted to get over the Modernist impasse, or failure to
> communicate with their users, had to use a partly comprehensible language,
> a local and traditional symbolism. But they also had to communicate with
> their peers and use a current technology.[12]

Over the past decade cruise passengers have been offered an ever-increasing
choice of vessels, with the result that operating companies have sought con-
stantly to upgrade facilities on board. Designers have raided history of all periods
and technology of all kinds to create onboard leisure environments of unparalleled variety and detail, environments which exactly fit Jencks's term 'ersatz or
artificial cities':

> There had to be a popular, consumer reaction to modern architecture and
> ersatz is it. What the modern architect would not do with historical styles,
> ambience and mood, the speculator has been happy to do.[13]

In this context cruise liner vessels may constitute a dimension of public architecture uniquely attuned to 'popular, consumer reaction', but one which also exploits
redefinitions and reworkings of a traditional and generally familiar visual vocabu-
lary. They have to be so attuned, since even though passengers cannot escape a
ship they dislike for the duration of the voyage, they certainly will not
re-book for another. As John McNeece, interior designer for several leading shipping companies, comments: 'Passengers have to like the spaces, otherwise I
must re-evaluate my work'.[14]

Modern cruise vessels have often been described as floating hotels, 'self-
propelled up-market Benidorms' and 'the floating apartheid machines of
postmodern leisure',[15] but certainly an accurate image would be floating gambling resorts; most guests are out of a hotel during the day, but the need for a
cruise liner to provide all-day entertainment while at sea stretches fantasy to the
limit – and beyond. Farcus is well placed at Miami Beach, and is responsible for
creating the extraordinary range of vessel interiors for Carnival Cruises. He is
very clear about his aims: 'I design for escapism. I believe that a ship should
actually be a discovery process. It's for this reason that I appeal to a very broad
range of people'. Much of his interior design for Carnival Cruises is based on
themed spaces, and here inventiveness goes hand in hand with a minute attention to detail in realising such 'artificial cities':

> I believe that passengers will marvel over the fact that this lounge, which will
> resemble the grandest of land-based show rooms, is actually on a cruise
> ship. Our theme rooms are designed authentically. We don't just throw the
> place together without any thought for authenticity . . . For example, in Cleo-
> patra's Bar onboard Fantasy I actually engaged a number of Egyptologists
> to make sure that the hieroglyphics were exact down to the last detail.

There are insights to be derived from comparisons between such exhaustively
'historical' replicas in a seagoing leisure and entertainment space and more
general ways in which modern cruise vessels convey both status and modernity

at a popular level through evocation of earlier periods and styles. Farcus sees alertness to a new class of passenger as central to the shift in the 1960s from liners as transport to cruise liners as popular floating entertainment:

when these liners withered as liners, their cruising capabilities similarly withered. However, even during this reduced period, cruising still did not appeal to the majority of people. Marketing had not been envisioned for the appeal necessary to reach out to the numbers required to make luxury purpose-built cruise vessels possible . . . Because this appeal had to be applied to 'average' persons who do not travel with servants, or live privileged lives normally, it was necessary to provide a degree of privilege on board that could be appreciated by the passengers.

As an architect, Farcus's approach to design is logical and thorough: 'Our expertise and talents begin with the planning of a design, be it building or a ship'. With ships in particular, 'Precious little is left to chance. Mistakes and omissions lead to unwarranted costs which are continuous and generally growing, or worse, to indifference on the part of the passengers'. Such hard commercial considerations are part of all ship design, whether merchant or passenger, but Farcus is equally clear that for him ship design also includes an important element of personal satisfaction:

Of course for me any design is a personal matter. Underlying the philosophy . . . is my commitment to honesty in design. That is to say that anything which I design must be pleasing to myself. It must be new, challenging, and intellectual. Design for me is not safe, but rather art which is created for a certain purpose and which must function . . . My belief has always been that architectural design should be mainly an expression of the creative arts.

Gianfranco Bertaglia sees the ship design process partly as the 'revisiting' of earlier styles and solutions which produce an architectural environment featuring traditional design elements familiar to the public in a manner appropriate both to modern taste and to a vessel's carefully-designed and specific function.

The P&O superliner *Grand Princess* (to be launched in 1997 at Fincantieri's Venice shipyard) exemplifies an imaginative 'revisitation' of diverse styles, both nautical and other. At an early stage in design the decision was taken to give a distinctively high and vertical profile to the ship's stern structure, and the similarity of this general shape to early galleon sterns – even down to the detail of the delicately-pointed transom at the waterline – led to the idea of accentuating the upper corners by protruding devices which echoed the prominent stern lanterns and galleries on such galleons. A later decision, and one inspired by very different imagery, was to span these two vertical and protruding elements with a large horizontal 'spoiler' across the whole width of the stern, thus giving the ship's profile a strong visual sense of purposeful directionality and, by association with fast cars, style and stability; this sense is sharpened also by the liner's unusually sculpted bow form. From this visual conceit came the solution of housing the ship's discotheque within the 'spoiler', thus eliminating any noise problem in relation to other onboard facilities; twin access lifts will be housed in the vertical pods which support it. The overall shape of the projected liner is both elegant and functional, and the visual language in which the design is articulated, and within which it can be explained (as here) is one which is readily available to, and

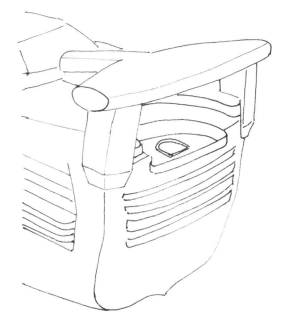

*Working drawing by Gianfranco Bertaglia of stern features on P&O's 1997* Grand Princess

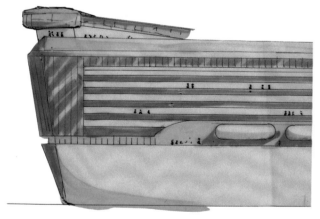
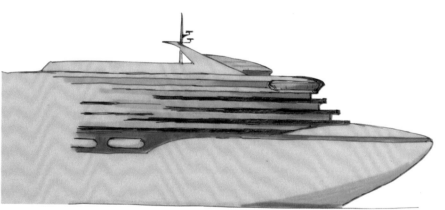

appreciated by, a general public interested in (say) modern production cars and pirate ships. In two planned Disney liners such elements are consciously foregrounded as part of a nostalgic 'adventure' package; in *Grand Princess* they constitute post-modern elements in an overall design which is unashamedly, and successfully, popular.

Through this cumulative series of informed and imaginative decisions, a design inspired by sources as distant as sixteenth-century galleons and the Ferraris of Maranello has produced a solution which is visually elegant, distinctive and completely practical. It is an example of modern architecture at work which would surely have pleased Robert Venturi, who in 1966 opposed what he saw as the 'puritanically moral language of orthodox Modern architecture' in his book *Complexity and Contradiction in Architecture*:

> I am for richness of meaning rather than clarity of meaning; for the implicit function as well as the explicit function. I prefer 'both-and' to 'either-or', black and white, and sometimes gray, to black or white. A valid architecture evokes many levels of meanings and combinations of focus: its space and its elements become readable and workable in several ways at once.[16]

As Jencks was to comment almost ten years later, 'if you want your architecture to "resonate with culture as a whole", then you'd better use a wide vernacular which includes all sorts of signs and traditional motifs'.[17] Modern cruise liners do precisely this, not to satisfy architectural theorists but, as Tillberg points out, because it is commercially important for passengers to feel proud of the vessel on which they are cruising when comparing it from ashore with passengers from other vessels. Competition is intense and a recognisable vessel profile is always an advantage. When several liners may be at anchor off any one Caribbean resort at a time, 'product identity' is crucial. One significant development since Tillberg's early work in the 1960s is that 'assignments today comprise total concept solutions. The function and design of both the ship's interior and exterior should reflect the operator's profile'.[18] Interiors of vessels are often coordinated across the fleet, with use of a 'reference ship' to set standards, while exterior recognition is today based less on the vessel's overall livery (almost all ships are white) than on distinctive funnel shape; the winged funnel designed by Farcus for Carnival Cruise Lines is truly an 'aero-nautical' icon which proclaims any ship that bears it to be modern, fun and part of a transport world which includes the jetliner. Functional it may be in dispersing exhaust gases to one side, but juxtaposition with an earlier, though equally functional tug's bridge makes only too clear its other role.

The design and development of passenger vessels, and especially of liners, has been written about extensively, but attention has centred largely on their interior decor, with scant appreciation of their exterior design, or of how this might relate to their function. Equally little has been written on the theoretical implications of earlier ship interior design which replicated in bizarre detail an architecture whose general proportions, finishing details and proportions had its bases in class-specific land-based structures; often a rather odd combination of delicate wooden latticework with massive girders and rivets. Modern ship interiors do not so much replicate interiors ashore as draw upon a wide range of reference to produce what is truly a free-floating and self-regarding interior world of spacious atriums, themed public spaces and enclosed sundecks.

*OPPOSITE, FROM ABOVE, L to R: The P&O superliner* Grand Princess, *designed by Gianfranco Bertaglia and to be launched at Fincantieri's Venice yard in 1997; Working drawing by Gianfranco Bertaglia of the stern profile for the* Grand Princess; *Working drawing by Gianfranco Bertaglia of bow profile and shape in relation to bridge form for the* Grand Princess; *ABOVE: Funnel of the 1996 Carnival Cruises liner* Inspiration *during fitting out at Helsinki, October 1995*

43

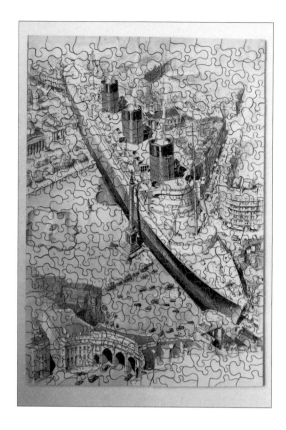

One advantage of the boom since the 1980s in building high-profile cruise vessels – a development which has centred on skilled European yards and has involved leading designers in several fields and countries – is that at least in this area ship design has achieved a new prominence. In earlier periods writing on ship architecture was undertaken largely by ship historians and enthusiasts whose interest was too often in mere technical detail and in reproducing (often in poor facsimile from contemporary journals) exactly those statistics of opulence which the liner companies themselves promoted. The motives of such companies were to impress upon potential customers the size (and therefore also safety) of their vessels by comparison with familiar city centre landmarks in London, Paris or New York. Much commentary since has relished the undoubted sumptuousness of liner interiors without speculating on how this might illuminate the aims of designers and responses of passengers or, indeed, how some of the visually bizarre advertisements and articles might have affected their readers' general responses to buildings and liners alike:

> the generation that turned forty in the 1930s had been brought up on popular magazines that had not only set discordant pictures next to each other (the significance of each being explained in a caption or text) but might in a drawing juxtapose the image of a skyscraper and a transatlantic steamer set upright on its stern to compare their relative lengths.[19]

Ship architecture has exerted a fascination on architects and the public alike, but, as mentioned earlier, has rarely been properly discussed and even more rarely approached with any alertness to a metaphorical or self-referential dimension. The London-based designer, John McNeece, says that he 'took a cutting from *Canberra* into *Oriana*'[20] in designing the latter's interior, while the fore and aft shape of Robert Tillberg's single funnel for the vessel also echoes *Canberra's* distinctive twin funnels, a feature she shared with some Shell tankers of the pe-

*ABOVE: 300-piece jigsaw of the Cunard liner* Queen Mary *(launched September 1934) by Chad Valley, late 1930s; BELOW: 'A fully postmodern ship', the Tillberg 'classical' profile of the projected* Queen of the Americas *superimposed on the 1951 United States liner* Constitution

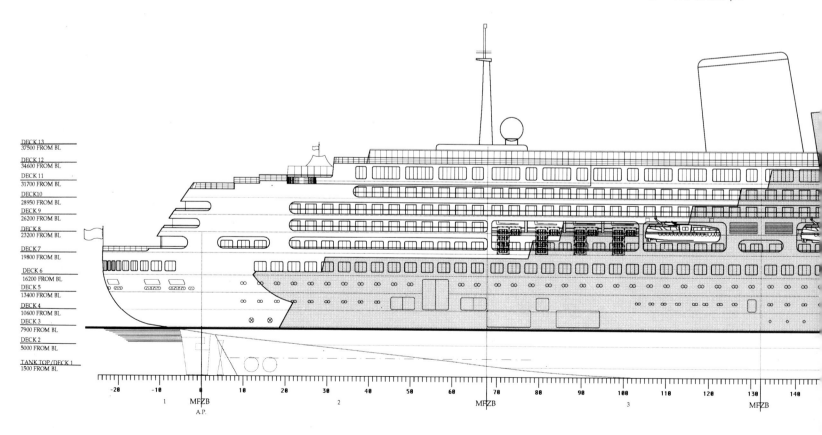

riod. Such specific examples of comparable design features are important in defining the shared elements of ship design.

Very different in nature and scope are two 'Classical Ocean Going Vessels' (the precise phrase in the project schedule) being designed by Tillberg Design for possible United States owners. Robert Tillberg defines the term with reference to this project:

> Classical design is expressed in many different ways. Key classical styling characteristics in her exterior lines are two large funnels, the long and open tiering of aft superstructure, the center of mass being forward of midships, a plumb superstructure front and strong horizontal layering.[21]

The Tillberg project brochure for the new liners emphasises such 'classical' echoes by superimposing the new vessel on the design of the old, in this case the 1951 SS *Constitution*. The brief specified this particular look for specific commercial reasons, so that the new vessels might blend in with the land architecture of ports to which they would cruise: 'An important part of the classical concept for the *Queen of the Americas* is that she is consistent with "Old Town" port districts in Hawaii and Alaska.

Such an explicit design relationship between ship and land architecture is rare (the design and coordinated paintwork of the original Star ferries and their terminals in Hong Kong are good examples), but it is clear from the Disney, P&O and this Tillberg project that cruise vessel owners are now looking quite consciously to 'quote' elements of design from earlier, and in some respects quite different, periods and examples of ship construction and use. As the Tillberg brief for the new liner states: '*The Queen of the Americas* will differentiate clearly from the current fleet of cruise ships and presents a highly functional, producible arrangement. She is a Post-Modern ship'. Especially interesting is the extent to which the designers entrusted with realising these visually challenging (and financially

*Costa Cruise Lines vessels at the Italian port of Genoa*

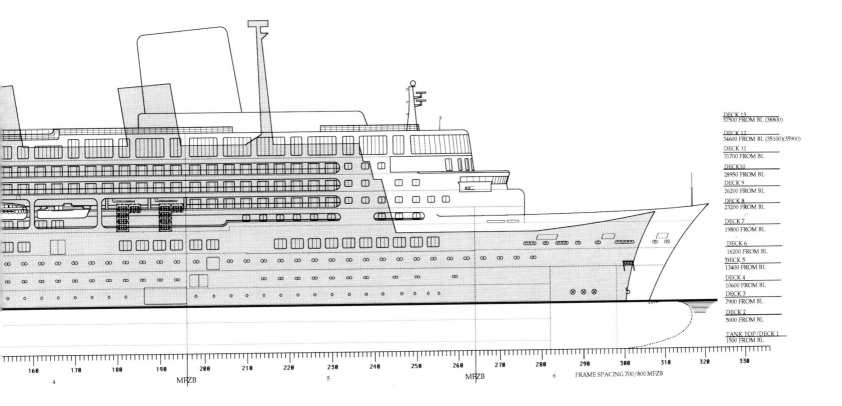

FROM ABOVE: Liner in a landscape – the 1995 Carnival Cruises liner Imagination at Miami; Tourist ship in tourist setting – the 1984 P&O Princess Cruises liner Royal Princess at Venice; OPPOSITE, FROM ABOVE: A ship is a world – 1928 Empire Marketing Board poster, now in the Public Record Office, 'Suez Canal' by Charles Pears; 'Setting sail', the 1950 P&O liner Chusan leaves under gloomy Southampton skies for a Mediterranean cruise

daunting) enterprises feel able to assume a certain nautical familiarity on the part of the public; for it is the public, after all, who will book cabins – or not. The difference between this situation and that of a land architect designing and constructing a high profile building for a corporate client could not be more clear. As Farcus puts it, 'it has become possible to design cruise ships for the cruising experience which appeals to a very broad segment of the population . . . For me this resulted in what I call entertainment architecture'.

The two large 'retro liners' for The Disney Corporation, a powerful new player in the lucrative and still-growing cruise ship business, also illustrate the selective 'revisitation' of earlier ship styles to offer a product which will have a popular, yet distinctive, appeal. The Disney vessels have been ordered from Fincantieri against strong competition, and are a new departure in passenger ship design with their clear post-modern handling of definitive features from both interwar transatlantic liners and the era of sail: twin funnels (the forward one is an observation lounge), oversize 'porthole' windows, 'classic' colours of red, white and black, a clipper bow and distinctive stern (with gilded decoration on both), and deliberately exposed anchors.

The richness of ship-architectural readings offered by such new vessels lies partly in the unpredictability of connections made by passengers. In this respect the combined mobility and reassurance of large ships offer unique opportunities for metaphors at once complex and unexpected, with chronology juxtaposed or suspended: on the Carnival liner *Inspiration*, Farcus designed the library in 'Elizabethan style . . . complete with Gothic arches, columns, rusticated walls, and coffered ceilings', while on *Ecstasy* 'really old-style neon signs' are woven into a motif which can 'transport the passenger' to a period seen as 'bygone, yet fondly recalled'. Such creations parallel the increasingly self-referential and self-contained worlds of the shopping mall, as Margaret Crawford suggests in her account of the West Edmonton Mall, which according to the *Guinness Book of Records* is the largest in the world:

> the mall presents a dizzying spectacle of attractions and diversions: a replica of Columbus's *Santa Maria* floats in an artificial lagoon, where real submarines move through an impossible seascape of imported coral and plastic seaweed inhabited by live penguins and electronically controlled rubber sharks . . .[22]

Crawford notes that such a mall needs only 'the addition of housing . . . to become fully habitable, a world complete in itself'. A liner provides this addition, if only for the duration of the cruise; in this context Joan Didion's observation that in themed shopping malls 'one moves for a while in an aqueous suspension' seems all too apt.[23] British imperialism also defined its own world, and central to the ordering imperial consciousness were 'the sea routes binding the Empire together',[24] while the public imagination could also thrill to tales of daring involving seafaring, shipwreck and survival:

> metaphors also related colonial writings genealogically the one to the other. For example, the paradigm of desperate and glorious adventure on board ship (also a parable of white exploration of the unknown) links *Westward Ho!* (1855) by Charles Kingsley to Captain Marryat's rather earlier popular sea fiction for boys, *Mr Midshipman Easy* (1836) and *Masterman Ready*,

and to Stevenson's adventure fiction. The same set of motifs reappeared, sometimes in ironized form, in Conrad's tales of mariners' solidarity and crisis at sea.[25]

Such motifs also appear, albeit in a minor key, in EM Forster's novel *A Passage to India*, published in 1924 but begun before the War, an absorbing study of British India by one who had lived and worked there as the private secretary to a Maharajah. In it the young woman, Adela Quested, 'thought of the young men and women who had come out before her, P. & O. full after P. & O. full'.[26] Significantly, the only British person in the book who responds imaginatively to India, Mrs Moore, dies on board ship while returning to England, and is buried at sea.

A key moment in the journey to India was the transit of the Suez Canal, that passage through the sands of the 'Middle East' by which vessels approached the 'Far East' – both Eurocentrically defined terms. One poster in a series by Charles Pears, entitled *The Empire's Highway to India* (a reassuringly non-nautical term), depicts the view from the ship's bridge as it passes through the canal, emphasising the narrow 'highway' straight ahead, with desert and camel trains either side; not until arrival at Bombay's Gateway to India dock will the British passenger have to encounter foreign territory or people. And to arrive in India as an imperial administrator was, supposedly, also to be 'at home'.

In earlier periods to sail to foreign parts was to experience adventure in different, and possibly dangerous, cultures; it was also relatively cheap, as Alec Waugh remarked of his first round-the-world trip in 1926:

When people used to ask me how I could afford to do so much travelling, I replied that I spent far less abroad than in a London flat. Travel, by ocean liner, was indeed in those days one of the cheapest ways of living, certainly for a freelance writer who carried his office with him and could work in his cabin or saloon.[27]

Today cruising is all too predictable, with passengers saved (maybe) from possible boredom by exotically themed public areas which offer versions of cultural or temporal travel in themselves. Another change is that society in many countries is now itself multicultural, and 'abroad' in some senses all too close to home. As Evelyn Waugh lamented conservatively to his diary in 1963: 'It was fun thirty-five years ago to travel far and in great discomfort to meet people whose entire conception of life and manner of expression was alien. Now one only has to leave one's gates'.[28]

The paintings and drawings of the marine artist and ethnographer Peter Anson, who lived and worked on the East Coast of Scotland between the wars, record an altogether more intimate working and visual relationship between land and sea culture, and in describing his own cottage quayside at Macduff in 1944 he captures that sense of a working port on a human and accessible scale that is now all but lost with increasing industrialisation in scale and process – albeit for better working and living conditions than those he records:

it would be difficult to find any house, no matter where you went, so close to shipping as is Harbour Head. Colliers unload their grimy cargoes at the very door. Cement-laden vessels cover the windows with fine grey dust when the wind is blowing from the west. Not far off, the sound of shipbuilding can be made out; mingled with the harsh cry of innumerable gulls; the hissing of

*The 1961 Shell tanker* Serenia *shared some design features with P&O's* Canberra, *one of whose designers had previously worked with Shell. These pictures of her on trials illustrate the elegant design and careful paintwork*

steam and the creaking of derricks on coasters, together with the quite distinctive music of different types of motor engines in the fishing boats.[29] Such easy proximity to the construction and operation of ships mostly vanished with progressive mechanisation of cargo handling and the need for secure bulk storage of increasing amounts of ever more valuable cargo. That trend has intensified, and with developments both in technology and scale the earlier and more intimate relationship between land and ship architecture has largely disappeared in most Western countries. What has replaced this relationship is an internationally mechanised scene in which shipowners commission elements of a single vessel from international designers and constructors. Cargo containers in Rotterdam's Europort are now handled by robot tractors: 'Watch out', the engineer who designed the system warned Sekula, 'they don't see you!' [30]

Some merchant vessels may never have a 'home port', their crew nationality as well as their port of registry determined largely by commercial considerations:

A scratchy recording of the Norwegian national anthem blares out from a loudspeaker at the Sailors' Church on the bluff above the channel. The container ship being greeted flies a Bahamian flag of convenience. It was built by Koreans laboring long hours in the giant shipyards of Ulsan. The crew, underpaid and overworked, could be Honduran or Filipino. Only the captain hears a familiar melody.[31]

In land architecture there is no equivalent for such complete internationality, not only of conception and construction but also of operation; a building puts down roots in many senses when it is planned and built, and when it pays its local taxes. Foster's Hongkong and Shanghai Bank building famously accommodates local traditions in its alignment and its colour; for all the international expertise in its building and use it is in a crucial sense very Hong Kong. Ships float free of the constraints applicable to land architecture, by which buildings are tethered to their setting by a complex network of wires and pipes, as well as bound in other senses by myriad laws and regulations that define them as an integral part of a specific location and administration. When a ship sets sail – a phrase which magically survives the technology – she must operate for the duration of the voyage as a completely separate social and economic unit, with its own strict hierarchy of command and its own risks, costs and rewards.

There has been a progressive divergence between ship and land architecture in basic materials of construction. The closest and most enduring similarity was between the simple wooden constructions of rural communities and their fishing vessels, while the most evocative image of shared shelter on land and sea is that of the upturned boat as dwelling. Many cottages in Devon and Cornwall use former boat timbers, some taken from wrecks, in their roof construction while the application of shipbuilding skills to land structures can be clearly seen in the timber structures on Bergen's historic dockside.

With industrialisation came increasing use of iron and steel, both in framing buildings (with increased height from the 1860s) and in building ships: the technique of cladding steel frames with various skins, including bizarre 'stone' facing panels as on London's Canary Wharf (a technique brutally revealed by the recent bomb at the Docklands South Quays station), continues in contemporary land architecture, but most ships now have an appearance – and not only in their hull

– which reveals their construction from sheets of relatively thin metal over a supporting frame. The surface of earlier vessels was explicit with its symmetrical rows of rivets, and welding brings its own patterns of stretched metal skin. Superstructure units sometimes have their reinforcement on the outside, while the hull's combination of weld shrinkage during building and sea pressure in service (especially at the bow, where waves pound, or along sides where vessels rub when docking) often results in a ship's skeleton being clearly visible, giving a 'starved horse' appearance. When the liner *United States* took the Blue Riband for the fastest Atlantic crossing in 1952, she arrived in New York with her bow section abraded almost to bare metal by the waves.

Although architectural writing is mostly silent on ships, scope for considering them as architecture is offered, albeit implicitly, by Reyner Banham's *Theory and Design in the First Machine Age* (1960) which contrasts a dominant 'compositional' or 'particulate' concept of architecture, in which 'small structural and functional members . . . are assembled to make whole buildings', with the approach of architects such as Mies van der Rohe or Skidmore, Owings and Merrill, and of engineers such as Buckminster Fuller. These designers worked, Banham argues, 'by subdividing a bulk volume to create functional spaces out of it', a procedure which accurately describes how the overall volume of a vessel, itself often pre-determined by overall operational parameters of draught and beam, is then apportioned by the naval architect between cargo or passenger space, propulsion unit, fuel, crew accommodation and stores. Questioned about the most severe constraint on his interior design work for cruise ships, Tillberg replied:

> The biggest restriction is the length of the fire zones as most of the cruise ships today are Panamax size, ie 33m in width, and the longest fire zones are approximately 45m. Therefore we never have more than 1485 square metres to work with.[32]

'Panamax' is shorthand for the maximum size of vessel that can use the Panama Canal ('Suezmax' is similarly used) and both passenger and cargo vessels work to exploit these precise dimensions to the limit; Panamax container vessels designed by Marshall Meek in the 1960s adopted a completely square deck space at the stern to enable the maximum number of containers to be stowed, and the box-like sterns of many cruise vessels are similarly designed more with the Panama Canal than aesthetics in mind. With container vessels there is also the commercial requirement to carry containers as far forward as possible, especially on deck, and when Meek was designing the first generation of such ships his model for the relation between deck and hull forward was not another cargo vessel but the aircraft carrier *Eagle*.[33] However, these early container vessels had such a pronounced flare that they tended to 'slam' in heavy seas, in later vessels the design was modified; compromise in ship design is always between the differing priorities of the market and of the sea.

The division of 'bulk volume' that Banham advocated is seen by him as 'rare' in architecture of the period he discusses (1925-1970), but he maintains:

> it may be taken as a general characteristic of the progressive architecture of the early twentieth century that it was conceived in terms of a separate and defined volume for each separate and defined function, and composed in such a way that this separation and definition was made plain.[34]

*FROM ABOVE: Ship carpentry – workboat, River Exe, 1995; Land carpentry –historic wharfside buildings, Bergen, 1995; OPPOSITE, FROM ABOVE, L to R: General arrangement of a refrigerated cargo vessel by Skipskonsulent AS; Transverse section of container arrangement by Skipskonsulent AS; Superstructure of the 1955 SS Shieldhall, Southampton, 1995*

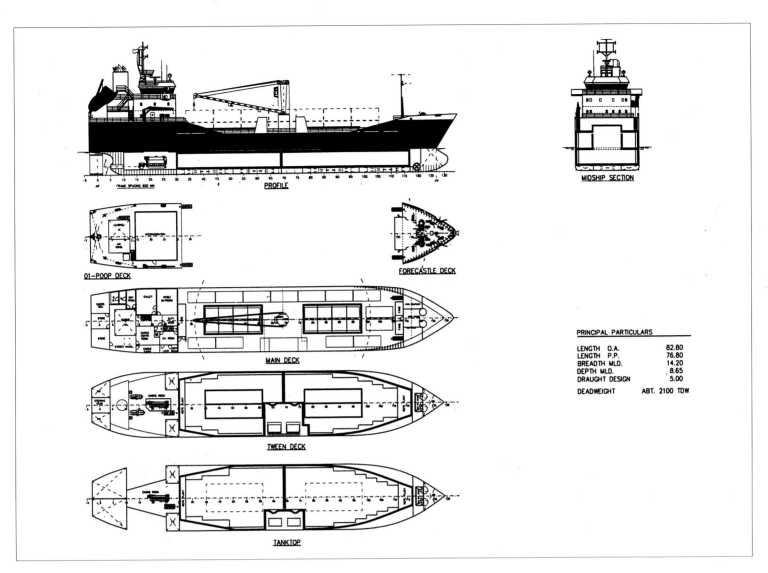

PROFILE

MIDSHIP SECTION

01-POOP DECK

FORECASTLE DECK

MAIN DECK

TWEEN DECK

TANKTOP

PRINCIPAL PARTICULARS

| | |
|---|---|
| LENGTH O.A. | 82.80 |
| LENGTH P.P. | 76.80 |
| BREADTH MLD. | 14.20 |
| DEPTH MLD. | 8.65 |
| DRAUGHT DESIGN | 5.00 |
| DEADWEIGHT | ABT. 2100 TDW |

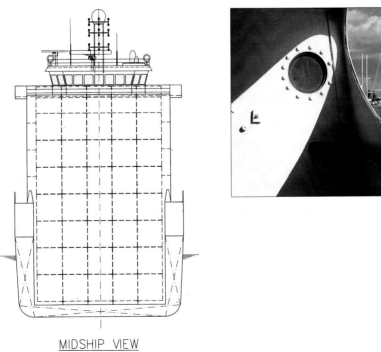

MIDSHIP VIEW

FROM ABOVE: *The wrinkles of welding* – Silja Serenade *at Helsinki, 1995; The Channel pilot climbs the riveted plates of the Alfred Holt cargo vessel* Aeneas, *Torbay, 1961*

On this criterion, ship architecture is decidedly 'progressive', but understanding ships as progressive architecture demands that we look at more than their appearance. The possible relevance of ship to land architecture is too often discussed at the level of spotting 'ventilators' or 'portholes' on buildings by Sir James Stirling and then referring to his father's time at sea as an engineer, or to Stirling's own studies at the port city of Liverpool. It is true that at the time (1945-50) when Stirling studied at Liverpool's School of Architecture – then the largest in Britain – both the active maritime life of the city itself and the policy of the Head of the School of Architecture, Sir Charles Reilly, encouraged links with the States.[35] Stirling himself once provocatively compared his controversial Engineering Building at Leicester University (completed in 1963) with 'the profile of an aircraft carrier, its bridge set to one side of its split-level flat-top flight deck'.[36] The difference between this aircraft carrier comparison and that drawn by Meek is revealing; one is based on appearance, the other on specific functional affinities.

Stirling has been described as an architect 'whose enthusiasm for the machine has generally been limited to an appreciation of the nostalgic power of the abandoned, technologically outmoded machines of the nineteenth century, such as the steamship'.[37] Fun though these elements often are visually, 'the deployment of toy-like evocations of ships' funnels'[38] – since widely quoted in other buildings – have little relevance to contemporary shipping technology. It is noteworthy that Disney have emphasised these elements in their 'retro-liner'. A more important aspect of Stirling's architectural awareness is his early interest in 'the industrial vernacular, in particular the anonymous but forceful nineteenth-century buildings of Liverpool'. In 1957 Stirling wrote in *The Architects' Yearbook*:

It should be noted that the outside appearance of these buildings is an efficient expression of their specific function, whereas, today, they may be appreciated picturesquely, and possibly utilized arbitrarily.[39]

This interest in constructions whose 'outside appearance' is an 'efficient expression of their specific function', not to mention subsequent 'picturesque appreciation', suggests fruitful comparisons with ship architecture. The 'anonymity' of many nineteenth-century warehouses also parallels the modest teamwork behind traditional cargo ship design; severe in style, such buildings hardly prefigured work by Rogers and others which has been seen as 'concentrating on the structure, looking for meaning in the often over-muscular display of the means by which the building lives and breathes'.[40] Like ships, however, these vernacular buildings did reflect their function in storage and transportation by their appearance, and the 'beauty achieved has never been the deliberate aim of the designers, most of whom have remained unknown and unsung'.[41] The novelist Virginia Woolf wrote similarly in the thirties of London's docks:

the aptness of everything to its purpose, the forethought and readiness which have provided for every process, come, as if by the back door, to provide that element of beauty which nobody in the Docks has ever given half a second of thought to. The warehouse is perfectly fit to be a warehouse; the crane to be a crane. Hence beauty begins to steal in.[42]

The internal spaces of such industrial buildings replicated in function the holds of the vessels that served them, and the labour-intensive stowage so necessary in both was rendered obsolete in the 1960s by standard containers in which goods travelled the world door-to-door in their own private space: secure, weatherproof and mobile. Appropriately, Stirling's early interest in Liverpool's maritime-related architecture itself found eventual scope in his 1984 commission to adapt the city's Albert Dock as an outstation for London's Tate Gallery, alongside the Merseyside Maritime Museum. In London, too, the cobbled lane (complete with 'traditional' ice cream vendor) behind what was once the industrial warehouse structure of Butler's Wharf is now part of London's tourist attractions: the structures of wholesale capitalism are now retailed as heritage. The Thames barge, once a common working boat, is now also a rare and restored element of that same industrial past; something to be photographed.

*ABOVE: From wholesale to retail – ice cream for tourists on the cobbles behind Butler's Wharf on the Thames, 1995; BELOW: A restored and converted Thames sailing barge passes Butler's Wharf, on which a sign advertises 'retail and residential accommodation', the Design Museum is on the left of the picture, taken from Tower Bridge, August 1995; LEFT: The 1977 gas-turbine fast ferry* Finnjet, *which operates between Helsinki and Travemunde, as built. Her bow and superstructure have since been remodelled*

# POPULAR TRADITION:
## *DYNAMIC CONTINUITY IN SHIP DESIGN*

*A large cruising liner is the nearest thing so far to a completely man-made total environment. It houses every kind of human concern, work, play, health, sickness, birth and death, and at its best the scope of its facilities transcends most buildings and many towns.*

*Kenneth Agnew, 'The Building of the QE2',* Architectural Review, *1969*

*I design for escapism. I believe that a ship should be part of a discovery process. It's for this reason that I appeal to a very broad range of people.*

*Joseph Farcus, architect 1995*

One consequence of the general neglect of ships in the field of architecture and design is that even famous passenger liners with a distinctive appearance are perceived as 'anonymous': the outstanding vessels built in Italy between 1951 and 1963 included the ill-fated *Andrea Doria*, the outstanding twin-funnelled sisters of the same year, *Raffaello* and *Michelangelo*, and the 1966 *Eugenio Costa* for Costa Lines. All these vessels were the work of one designer, Nicolo Costanzi, already a noted ship designer when he planned the Italian cruise liner *Victoria* in 1931, but his name remains virtually unknown today.[1]

Such anonymity is regrettable, as the impact of the designer is crucial. In his 1993 autobiography the late James Gardner, design coordinator for the exterior of the *QE2*, recalls how he was first asked by Cunard in the 1950s to shape their projected new transatlantic liner, the *Q3*. He especially remembered entering their Liverpool design office:

> I climb up the narrow stair to the Cunard drawing office to find oak plan chests, high wooden stools and pale-faced draughtsmen with a problem. This ship will have no sweeping curves: she is a block of utility flats dumped in the sea, and must ride uncomfortably high to pack in the essential accommodation – more like a piece of floating real estate. A clumsy model, bits borrowed from the old *Queen Mary* in an attempt to make her look like a ship . . . I am itching to have a go.[2]

Gardner effected remarkable changes in concept and design, not least elimination of the sacred Cunard funnel, which the company was keen to keep, but which would have produced a very strange liner indeed, as photographs of a contemporary model demonstrate. Plans for this Cunard transatlantic vessel were eventually abandoned by the company in 1961, but Gardner went on to design what became the *Queen Elizabeth 2*, a ship planned for greater operational flexibility by cruising during the winter, which was launched in 1969 and owed much to Gardner's initial development work on the earlier project.

Colour profile studies for the original *Q3* project which are still in Gardner's

*OPPOSITE: The 1993* Kong Harald, *a Hurtigruten passenger ferry designed by Skipskonsulent AS; ABOVE: Photograph of in-house model for the Cunard Q3 project*

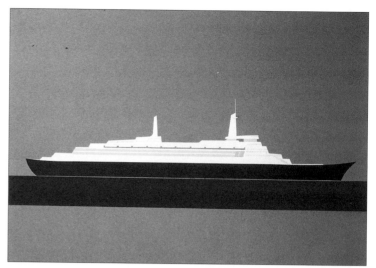
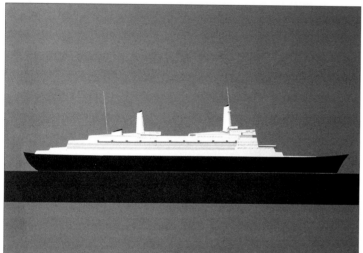
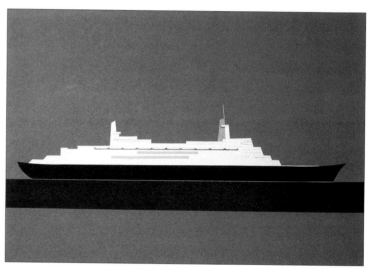

*Studies by James Gardner for the Q3*

London studio reveal not only the attention he paid to the overall appearance of the vessel, but also his understanding of key decisions such as where to locate the funnel – or funnels – and what profile, height and colour they should be. The differing designs are all based on a standard hull, but the character of each is defined by the profile of superstructure elements, and by the relationship between these and various funnel arrangements. In his autobiography Gardner dismisses these drawings as 'designed to impress the press', but they are too careful for that; this is graphic work in which the designer seeks to satisfy himself. In retrospect Gardner was also typically self-deprecating about the whole transatlantic project, which at the time must have seemed a complete dead end:

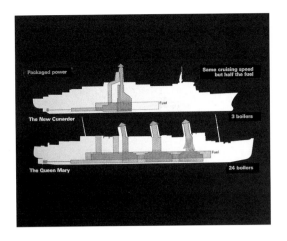

> I had just taken part in the greatest non-event in shipbuilding history. The *Q3*, as my mythical monster had been named, was dead as a duck and a chance to follow in the footsteps of Brunel was lost.[3]

Other drawings by Gardner illustrate considerations influencing his presentation of the new liner in the wake, as it were, of the two *Queens*: the size of the engines, the need for ventilation and the proportion of open deck space. As ever with ships, design on this project was very much a question of satisfying specific functional requirements, as well as of producing a good-looking vessel. And Cunard and Gardner were in agreement, despite their differences, that the latter aim was essential to the commercial success of the liner and a crucial function of whatever design was adopted.

As Gardner was very aware, the *QE2* was built at a pivotal period in the design and construction of large passenger vessels; like the 1961 *France*, now extensively remodelled and cruising as the *Norway*, she evoked an earlier period of ocean travel. Some might feel that Gardner's critical description of Cunard's initial design efforts on the abortive transatlantic ship – 'a block of utility flats dumped in the sea' – sums up many liners built since the *QE2*, and certainly the profile of liners has changed considerably since that of the 'ocean greyhounds' that competed across the Atlantic. Speed on scheduled runs was then a selling point for passengers, whereas today cruise vessels vie not in speed but in luxury, interior decor and onboard entertainment.

Gardner had never designed a ship before, and it is a measure of his professionalism that the vessel he produced was such a success. He was understandably somewhat daunted by the scale of the undertaking, and in a wry sketch depicts himself on the upper decks of the ship, then building on Clydebank, together with the chief naval architect Dan Wallace. Especially chastening was to see his studio drawings taking three dimensional shape on such a scale: 'Delicate lines drawn with a 2H pencil became muscular men fighting the stiff resistance of steel'.[4] The final impossibility seemed the launch, equivalent to any architect seeing a half-finished building released to slip at some speed into the nearby river. Gardner doubted it would, or could, really happen:

> Can that massive chunk of landscape move? No. After an interminable period of suspense (I am wondering whether we had better just go home and forget the whole thing) there is a sharp crack, a gathering roar and, through a cloud of red-dust from the drag chains, she sends a wave down the Clyde as though a glacier had calved an iceberg. And there we are, looking at an empty space.[5]

*Comparison of the 1934* Queen Mary *and the new* QE2 *comparing engine space as percentage of overall volume*

*FROM ABOVE: Cartoon by James Gardner of the QE2 on the stocks; Working sketch by James Gardner for Q3 bridge front; The classic Cunarder Mauretania, 1939, here at Southampton, illustrates the contrast between dark hull and the layering of shapely superstructure, set off by grand funnels*

In 1967 Gardner described how he sought to disguise the straight lines of the *QE2*, this 'rectangular box (economy) with no sheer, with deck angled up fore and aft'.[6] The result was, as he put it, that 'conscious design' went into the project in order 'to achieve acceptable lines'.[7] He wrote also of the design compromises dictated by financial constraints:

> The structure was basically rectilinear, as we can no longer afford the sweep and camber of the old 'craft built' ships, owing to labour costs. I, therefore, had to camouflage the rigid structure to give an effect of sweeping lines. In result, she is probably the last ship to be 'Savile Row' tailored, as against the 'Burton' economy tailoring now current in shipyards.[8]

Within these financial constraints Gardner made bold decisions on the appearance and function of the vessel, many of them as much to do with its necessary character as with any narrow definition of functional efficiency. A rare glimpse of a ship architect making exactly this case comes in a letter from Gardner to Commander Wood at the Cunard Head Office in Liverpool in which several quite drastic design decisions of the *Q3* project were summarised with characteristic pithiness:

> Major changes involved in the layout of upper decks: in my view the wheelhouse should be exploited to give the ship personality – and I have raised it one deck. The wheelhouse top is not open to passengers, but for navigational use only . . .
>
> The aft outlet vent, if placed immediately over the boiler casing, would be too close to the front one to look right, and so I have set it as far back as I think is possible without disturbing main accommodation layout . . .
>
> The aft vent requires some deck superstructure at its base (a) to mask the new line of the vent ducts and (b) to give it a proper visual relation to the build-up of the ship's profile.[9]

These key decisions are not only those least explained, but least explicable: Gardner felt the wheelhouse should be 'exploited'; a vent would not 'look right' in a certain position but needed 'a proper visual relation to the build-up of the ship's profile'. He later reflected on the blend of conscious design and partly instinctive tradition that had shaped earlier Cunard liners, noting that 'no ship can be considered a design problem in isolation' since ships 'have evolved to a formula and discipline of their own'. Gardner felt that 'within this pattern' the new liner had a special place in both continuing the image of 'the dashing Atlantic Cunarder' while facing wholly new commercial demands, such as the planned schedule of winter cruising.[10]

Clear in his own mind that the grace of much-admired Cunarders such as the 1906 *Mauretania* was 'a natural result of the shipbuilding practices of the time', rather than a conscious 'design project' in the modern sense of that term, Gardner sought such understatement in his own project, despite having to employ 'conscious design'. It was a classic case of art concealing art:

> A sweeping sheer line, low superstructure, characteristic vent cowls and tall smoke stacks added together produced, with a little attention to symmetry and tidiness, a beautiful ship . . . no visual designer, no gimmicks.[11]

Gardner saw the *Queen Mary*, launched in 1934 and now moored at Long Beach, California, as already 'more hotel and less ship', a shift which had entailed 'care [being] put into the architectural detail of her deck houses, window layout and bridgefront' so that 'What was lost in grace was gained in scale'.[12] He felt the design of the new *QE2* was complicated by the fact that as a liner she was not only 'outscaled by the 200,000-ton tankers' but was also being built for an age in which technological admiration had moved elsewhere: 'the bigness of ships no longer impresses a people who take supersonic flying and space-craft coolly'.[13] Gardner argued, interestingly in the light of some subsequent developments in passenger ship design:

> We are more interested in the latest fashion-styled car. A smoothly stream-lined Cunarder with raking mast would please small boys and the publicity people. Nevertheless, when considering her profile I found it necessary to lean over backwards to avoid that look which we associate with product styling. *She is already functionally streamlined . . . under the water* [14] [my emphasis].

This is a refreshingly straightforward expression of conceptually and practically complex design issues, and Gardner is alert to the range of diverse demands which would determine the final shape of his ship (later priorities would, to his intense annoyance, again change its 'final' shape). Like David Pye he is also clear that simplistic appeals to 'the functional' do little to resolve such issues:

> Contrary to theory, purely functional forms only look right when they are shaped by very special conditions: a billiard ball and a submarine are cases in point. A large passenger liner adapted for cruising as an assembly of many parts with different kinds of function would, without some visual control, end up as a happening.[15]

As modern technology 'replaces rule-of-thumb craftsmanship' the central element of what Gardner terms 'visual control' becomes ever more necessary, especially if it is accepted that part of a good designer's job 'is to please the eye'.[16] When even large ships were still being designed by 'rule-of-thumb', individual human judgement (what Bertaglia terms 'educated imagination') could intervene at each stage and level to compare hull with superstructure, funnel with bridge and masts, stem rake with sheer and flare. Modern technology – and Gardner was working without computer assisted design – can certainly devise efficient solutions to practical requirements, but there is still no substitute for an expert eye cast over the whole design so that, in Gardner's own phrase, everything should 'look right'. It was this attention to both detail and overall proportion that led to his being understandably outraged when more staterooms were added to the *QE2*'s upper accommodation deck, and her elegant (and functional) funnel altered. As a conscientious designer sensitive to the tradition of great liners in which the *QE2* took her place, Gardner inevitably felt that such alterations betrayed those very principles which, together with Cunard, he had sought to express with such care in the vessel's form.[17]

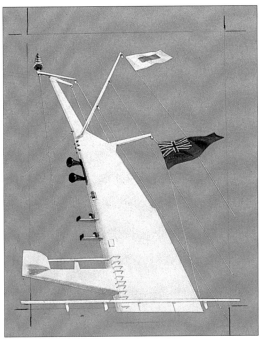

The original design of the *QE2* well exemplified that fine balance between innovation and restraint that Gardner saw as appropriate for a modern transatlantic liner in the Cunard tradition. In his autobiography he wittily describes the severely practical methods employed to arrive at a final superstructure profile for the liner, including the late night purchase and paring of a slab of cheddar cheese to attain the final curve of the bridge front. Photographs from the period record the

*The QE2 funnel by James Gardner as originally designed and constructed; Colour study by James Gardner for the QE2's foremast*

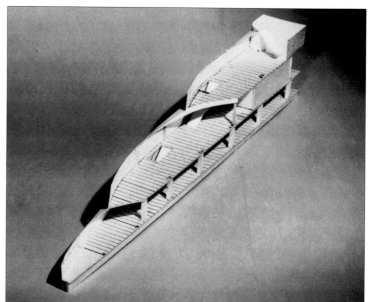

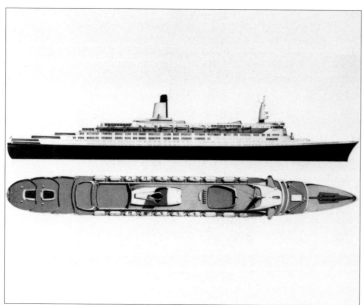

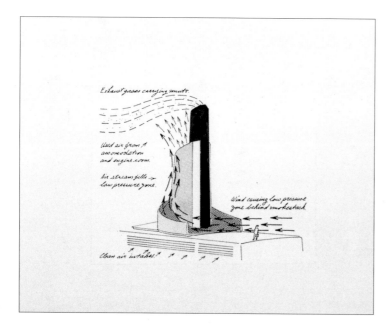

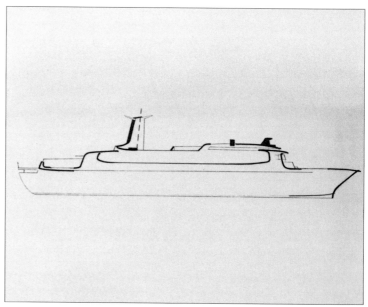

balsawood and card model used for designing the liner's stern decks, and it is hard now to believe that these finally became the elegant lines of the finished vessel.[18] It was this remarkable process of transforming pre-existing standard shapes into the distinctive and functioning unit of a ship that the architectural critic and photographer Eric de Maré had earlier sought to capture in his picture *Clydeside Scene*.

James Gardner was subsequently asked by Cunard to design *Cunard Adventurer*, the company's first liner entirely for cruising. This project entailed giving distinctive character to a largely pre-existing hull form, and studio photographs of early studies illustrate that designing a ship is no mere matter of assembling details, but also of ensuring a satisfying balance of mass, and of understanding the play of light and shadow across it, as well as striking a happy relationship between hull and superstructure. These photographs are different views of a rough model assembled in order to assess those basic proportions which define how a vessel looks on the water, and record progress from a bare model of the hull form alone to profile drawings and a fully-realised model shot in perspective. In his autobiography Gardner annotates the basic profile drawing of the *Cunard Adventurer* reproduced here, stressing the black raised bridge ('accent at nerve centre'), the unusual lean of the funnel ('raked forward to provide tension') and the shading of the superstructure behind the lifeboats, a technique he had also used on the *QE2* ('structure shadow-painted to link lifeboats').[19] Such aesthetic concerns are central to establishing a decisive and coherent personality for the vessel, a quality that some cruise vessels since have lacked as enclosed superstructure has grown and there has been correspondingly less scope for such wideranging and individualistic design features. Excellent designs continue to be built, as this book shows, but as overall size and bulk has increased, so the main accommodation unit of cruise liners has come to dominate both their profile and the punctuation of their surface, in some cases inviting visual analogies with the multiple-unit composition of container vessels.

As we have seen, the role of the 'architect designer' with regard to ships is problematic; Gardner was correct in seeing ship design, at least in Britain in the period preceding the *QE2*, as being essentially conservative and largely rule of thumb.[20] His own experience here is entertainingly instructive: when he encountered problems in designing an efficient funnel for the *QE2* (essential to prevent passengers being covered with smut) he sensibly enough sought information on both Victorian chimney pots and the very distinctive funnels already used on the liners *France* and *Michelangelo* – the former, he was told, were designed 'by "tradesmen", not PhD research engineers' and as for the latter, 'No one seemed to know, or care; foreign ships'.[21]

Such complacent insularity, together with a sense that some things are the concern not of designers but of 'tradesmen' – or at best engineers – still runs deep in many areas of British culture and the distinction it assumes between 'engineering' and 'architecture' we have already seen challenged by Foster. A certain caution is understandable in ship design for purely practical reasons: sailors and shipbuilders alike would probably agree with Robert Hughes's response to the architectural excesses of Brasilia and feel that 'it is better to recycle what exists, to avoid mortgaging a workable past to a non-existent Future, and to think small'.[22]

*OPPOSITE, FROM ABOVE, L to R: The 1971 cruise liner* Cunard Adventurer, *with exterior styling by James Gardner; Balsawood studio model by Gardner of rear deck structure for the QE2;* QE2 *profile and deck layout by Gardner; Studio model for* Cunard Adventurer *by Gardner; Drawing by Gardner of planned QE2 funnel showing forced air flow to lift exhaust smuts clear of decks; Working drawing by Gardner for establishing the overall profile of* Cunard Adventurer; *FROM ABOVE: Aerial photograph of the QE2's rear decks; This photograph by Eric de Maré, entitled* Clydeside Scene, *imaginatively illustrates the process and product of shipbuilding, from abstract shapes and volumes to finished vessel*

Revolutionary changes in cargo ship design since the introduction of containers in the 1950s – decades during which for the first time airline passengers equalled those carried by ship across the Atlantic – have fully demonstrated the ability of ship designers and engineers 'to get things done'.[23] Yet beyond the purely practical achievement there are analogies to be drawn between the attention that Foster devotes to achieving a smooth and tense surface to his structures and recent styles in liner construction; or between the aesthetics of buildings such as Lloyd's or the Hongkong and Shanghai Bank, both of which advertise their assembled nature (the latter is designed to be rearranged if need be), and the appearance and function of modern cargo vessels. Some of these are designed and launched with later 'jumboisation' in mind (or are sometimes even jumboised before completion, to respond to changing markets). The important difference, however, is that whereas some buildings proclaim their supposed technological sophistication, in ships such elements are solely functional:

> Both Foster and Rogers have lavished more attention on, and accorded more prominence to, the window-cleaning and maintenance systems than could be justified on narrowly functional grounds. Lloyd's has its blue tower crane gantries, the Hongkong and Shanghai Bank its maintenance cranes suggestive of star-ship gunports, both tributes to the idea of an unfinished process, with an appearance of being capable of adding to or extending itself.[24]

The Italian architect Renzo Piano had, like Gardner, never designed a ship when he was asked by Fincantieri to lend his name and style to the cruise liner *Crown Princess* (1990) the firm was completing for P&O. This vessel had been designed by Gianfranco Bertaglia, but as this was the first passenger ship built by Fincantieri for twenty-five years, a high-level management decision was taken to involve an already-famous land architect at a late stage. Only some thirty per cent of Piano's original design survived in the liner as eventually built (he had, for example, envisaged a fully-rounded 'dolphin' profile enveloping the whole vessel, which would have meant too great a loss of deckspace to be commercially viable). Much was made in subsequent publicity of Piano deriving inspiration from the dolphin for the shape of the vessel, and the one-piece bridge structure strongly retains this idea. The claim is credible, but more important are the final practicality and profitability of the ship. Chance inspiration for the liner's visible profile (Riccesi's 'passive' element) is here to be seen in the context of those 'active' factors which were translated into a functional and commercially successful ship: 'a thing must be beautiful as well as efficient if it is to be considered well designed: efficiency alone is not enough'.[25] Piano's intervention in the *Crown Princess* project is interesting for a number of reasons, not least the motive of executive decisions to have the vessel 'styled' so publicly. The *Crown Princess* as originally contracted for was a vessel less different certainly, but possessing the same distinctively Italian elegance as the other vessels which followed her from Fincantieri. Opinion will probably remain divided as to whether Piano's intervention improved the design or whether the dolphin inspiration was little more than a gimmick; the vessel itself remains an intriguing illustration of how land and ship architecture, and the status of those who design them, are regarded.

Piano's architectural interests took a different maritime turn in a controversial

*OPPOSITE, FROM ABOVE: P&O's 1990* Crown Princess *as launched; P&O's* Crown Princess *as at contract signature and designed by Gianfranco Bertaglia*

port development for his home city of Genoa, which in 1992 staged the 'Colombiadi International Expo' to mark the 500th anniversary of the arrival in America of its former citizen Cristoforo Colombo. Genoa has a proud seafaring history, and the ocean terminal for great liners such as the *Conte di Savoia* and the *Rex*, the liner which in 1933 took the Blue Riband of the Atlantic for Mussolini's regime, still stands – although in a largely abandoned state. In the postwar period industrial disputes and failure to adapt to container traffic accelerated Genoa's commercial decline as a port. Piano's bold redevelopment of part of the historic harbour area was intended to reunite city with harbour, and included a ship-shaped floating aquarium and exhibition space, the 'Nave Italia', as well as the 'Bigo', described as 'a system of masts, yards and cables, inspired by cargo ships, acting as the Expo's symbol and identification point'. The scheme was beset with economic problems and civic disputes, but its realisation demonstrated a rare recognition – admittedly in a specific maritime site and context – that the traditions of ship and land-based architecture might inform each other. The subsequent refurbishment of the 'Ponte dei Mille' port building, together with the construction of a new passenger terminal, may yet transform Genoa's fortunes as a port of call for cruise liners, and have more practical success than Piano's designs which, while inventive, seem not to have had a lasting use beyond 1992.[26]

Over the decades some well-known designers such as Bel Geddes, Raymond Loewy and, more recently, Piano himself have made much of how ships look, and certainly a ship's personality (a quality peculiar to marine constructions) is defined largely by appearance.[27] Yet the minimum requirement of any vessel is that it floats; buoyancy and stability must always override appearance. Debate on these issues is often confused because critics fail to address crucial differences between the 'active' and 'passive' elements of a vessel's form. Critical writing on the design of ships has been predominantly concerned with the latter rather than the former, whereas the informing concerns of naval architects are, as we have seen, necessarily the reverse.

A ship only earns profit when it is moving and loaded, yet even in harbour a ship is never still and constant attention is needed to mooring cables, fenders and gangways when loading or unloading, especially if weather and tides change, or other vessels manoeuvre nearby. Any unplanned contact with *terra firma* is 'going aground', the dread of all mariners, though if the bottom is sand and the weather calm, some vessels can survive well enough. The small coaster illustrated here can survive this temporary grounding since it is designed to bottom in small ports while working cargo. The only other exceptions are beach fishing boats, and the brief periods all vessels spend in dry dock for maintenance and repair. The shape of a commercial ship is a compromise between the profit motive to maximise use of interior volume and the need for a (largely submerged) hull form which is both safe and efficient to construct, operate and maintain. A ship which carries more cargo than its competitor but cannot sail in rough weather is not efficient; cruise passengers and corporations expect regularity of service. Conversely, a vessel which is superbly seaworthy but carries ten per cent less cargo than its competitor is not efficient either. Such calculations are delicate and vessel-specific, but their competing claims will determine the final shape that takes to sea.

*FROM ABOVE: The British coaster* Onward Mariner *aground at Knott End; Fishing boat, Dungeness; Wooden fishing boat, Sidmouth, 1995; OPPOSITE, FROM ABOVE: Fine lines – the elegant practicality of a modern Dutch trawler; Plans for modern Norwegian fishing vessels by Skipskonsulent AS, blend a traditionally elegant bow form with high-tech fishing gear. Owners and captains often wish to retain attractive bow lines from an earlier period and to have a bulbous bow as an expression of modernity*

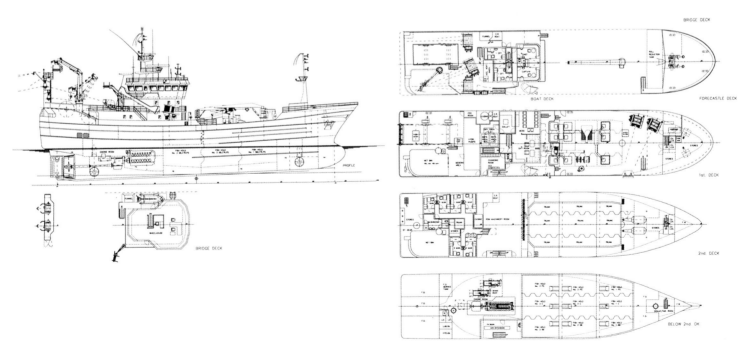

The 1960s were significant not only for the final demise of the passenger liner but also for the less publicised, and indeed less public, rise of the ocean-going container vessel. In worldwide passenger transport the silhouette of the Boeing 747 has replaced that of the ocean liner, while in the field of freight the container ship, with its distinctive multi-coloured cargo units, has become the truly expressive liner of our post-modern and market-driven times.

The cruise vessel and the container ship are the twin poles of modern ship architecture, not so much in contrast as in expressing elegantly efficient responses by the shipping industry to very different markets. Both are highly specialised and technologically sophisticated; both have produced astonishingly beautiful vessels which bear comparison with other landbased complexes devoted to the needs of the public – from leisure centres to supermarkets – and with the specialised infrastructure that services them. Container vessels are visibly an integrated part of a unitised, regular and global cargo web, while modern cruise liners cater for passengers who often fly to and from their port of embarkation; their sea voyage is but one part of today's on-screen and plastic-paid holiday booking. In passenger terms cruise vessels are by no means alone in the varied facilities they offer, and today even airports, far from merely providing rapid transit between flights, recognise the potential value of the customers they house in transit:

> In Britain, BAA [British Airports Authority] has earned enormous sums from franchising duty-free sales areas. It learned the habit of seeing things in retailing terms, and it started to replan its terminal buildings not for optimum passenger flow but to ensure the most prominent position for the duty-free shop.[28]

The London-based architect, Ian Ritchie, confirms this trend, commenting that 'today the strongest impression I have is that most major interchanges are now designed to make more money from retailing and advertising than they do from their "primary" function'.[29] Airports are static and cruise ships mobile, but in different ways both have their markets 'captive' to an extent that shopping malls can only dream of.

Such economic issues also affect more modest vessels: ferries on the lucrative English Channel routes have fought to retain valuable duty-free franchises against complaints from Euro-Tunnel that this gives them an unfair financial advantage in fierce cross-Channel competition. All modern ferries on this route now feature gaming machines, extensive duty-free shops, bars, restaurants, cinemas and live entertainment; on the other hand, finding a door to the limited (but free) open deckspace is a real challenge. In this respect, as in many others, such vessels seek to keep their passengers happy not by drawing attention to their seagoing aspects (as land-based architects and designers sometimes urge) but, as they have long done, by offering essentially land-like distractions: 'The opulent interior spaces of first-class travel on the steamships of the late Victorian and Edwardian periods may well have been designed to compensate for [absence of that] rich diet of changing tableaux' which the comforts of train travel could offer. This was especially so for those poorer passengers who were unable to afford 'the outward view from the promenade deck or its inner surrogate in the grand saloon'.[30]

Such vessels must seek to retain something of the ship's traditionally liberating image, while at the same time scrupulously avoiding imagery which might remind

*The 1990* Costa Marina *(built in 1969 as the container ship* Axel Johnson*) showing pool deck and three-funnel grouping*

their potential 'customers' that ships do actually plough the waves:

> the exaggerated jet-age streamlining of the cruise ship is a compensatory attempt to sustain that longing. The image of acceleration lends itself to idleness, to aimless floating pleasure, masking the monotony of the sea, the claustrophobia of the ship, and the driven life of short vacations.[31]

The whole issue of how the design of passenger space on such vessels relates to the encircling sea is problematic. On cruise ships designed for the American market, open decks are mostly limited; the ocean is to be viewed only through tinted glass, not encountered as anything approaching first hand. With its new liner *Oriana*, on the other hand, P&O sought to distinguish the ship from this market through extensive open decks and the use of natural wood and brass fittings with a 'traditional' feel; the Swedish studio Tillberg Design undertook research in consultation with the company to achieve this atmosphere also with the decor. This design decision, which deliberately seeks to evoke a sense of tradition relating to P&O's proud history since 1837, also makes sound commercial sense as the liner's home port is Southampton and most of her passengers will be British.

The high status of liners has always been matched by their high technology, with speed and comfort central to their function. The large boilers required in steam-driven vessels needed several funnels of considerable height for smuts to stand even a chance of clearing passenger decks, and from this severely practical origin funnels traditionally assumed importance as a key indicator of vessel size and status. On *Crown Princess*, by contrast, the single funnel is reduced to little more than a vestigial pipe to contrast with smooth superstructure lines, while other vessels have incorporated it almost entirely into their overall profile. Carnival Cruise Lines have the distinctively shaped and painted winged funnel, while a group of recent vessels for Costa Cruises have three slim funnels, not in line but elegantly grouped as free-standing columns towards the stern. Passenger companies have always been alert to how their vessels' appearance could influence what is now termed customer choice. In the 1920s motor propulsion was replacing steam and the Italian Cosulich Line, which operated transatlantic and other services, not only considered the effect of such technological changes on the appearance of its vessels but discussed the consequent aesthetic implications in a publicity leaflet:

> A problem which has occupied the shipping company and the naval architects considerably was that of the general profile of the ship and her superstructure. Aesthetically, no less than technically, the motorship represents a development which can offend against traditional lines. The new engine, heart of the ship, gives its own shape to the hull. The funnels, by whose number one used to deduce the power of the steamship, have lost their function and their significance.[32]

Although Le Corbusier had praised Cunard's 1913 *Aquitania* for the lessons its 'architecture' held for house designers, within ten years others were already making clear distinctions between ship and land architecture. The Italian architect, Edoardo Persico, wrote in 1933 that the country's new transatlantic liner *Conte di Savoia*, launched in Trieste the previous year, should be discussed by architects solely as an industrial product which had nothing to do with 'art'. His

comments were directed sharply at Le Corbusier's comparisons between the style of the *Aquitania* and the architecture of seaside villas:

> The position which we must take today, after a long moral polemic, is a rigorously aesthetic one: we do not believe, for example, that one of the decks of the *Conte di Savoia* unloaded onto a beach at Viareggio or Ostend, would be outstanding seaside architecture ['architettura di villa'].[33]

The prominence of passenger liners in design history, and the readiness with which Le Corbusier's superficial analogy between their superstructure form and that of land architecture has generally been accepted without question, makes them especially important historical examples through which to explore this distinction. Notwithstanding Persico's clear 1933 distinction between the function of ship architecture and that of a seaside villa, James Steele's recent book on the *Queen Mary* describes the verandah grill facade as being 'like a perfect fragment of a modernist villa'.[34] Such an approach does little to further understanding of ship architecture, and the double-spread photograph reproduced by Steele illustrates the extent to which the deck gear of the liner impinges on the design even of this relatively small area of superstructure.

The increasing focus of design criticism on interior architecture in passenger vessels should be considered historically, and specifically in relation to liners in which exterior form served a more varied function than the all-enveloping forms of today. Ships are a strongly unified entity, and from the moment a shipping company approaches a naval architect or a shipbuilder with a design brief, however detailed or vague, a ship is typically more specialised than any land building; relatively few ships are converted to new uses, though an important recent exception is the conversion at Genoa's Mariotti yard of two former United States container ships into passenger vessels for Costa Cruises. An earlier example was the 1966 conversion of the 1952 Swedish-built tanker *Soya Margareta* into the Greek roll-on roll-off vehicle and passenger ferry *Minos*, expertise in which Greek shipyards are unparalleled. The attention to style and detail that such successful conversions requires reinforces Donato Riccesi's complaint at the scant attention paid to ship architecture in professional debate:

> Perhaps it has been only a kind of mental laziness (the difficulty of encountering a new, technical, engineering-related terminology) that has meant that the field of naval architecture has been addressed only in a marginal way, remaining within the scope of interior design, considered separately from the hull that carries it.[35]

Riccesi also suggests that the uniqueness of most ships, the fact that they are not designed with series production in mind, is one reason for their neglect by writers on industrial design; by contrast 'the Vespa of D'Ascanio, the ETR 200 electric train of Ponti-Pagano and the FIAT 500 of Giacosa have been discussed exhaustively in most recent publications'.[36] Even The Design Museum in London, located in a former warehouse overlooking the Thames, offers only one ship item on its interactive visitor database (an early Meccano model of the *Mauritania*) although the usual Vespas, Citroen 2CVs and Italian coffee machines are displayed.

Yet some naval architects felt early the need for a distinctive and appropriate architectural language in ship design, and in Italy at least, such debate found both theoretical and practical expression. The Trieste architect Gustavo Pulitzer

Finali argued basic principles in an interview prior to the maiden voyage in June 1931 of the innovative motorliner *Victoria*, for which he designed the interior:

> Not architectures which superimpose themselves on those of the ship, not false palaces, not temporary structures. Architecture must seek its harmony in the spirit of the fittings, without altering the spaces which are offered by the structure of the ship itself. Countless pleasing effects can be achieved by studying the possibilities most appropriate, and at times also most intimate and hidden, that every material offers for decorative expression.[37]

It was this principle of 'harmony in . . . the structure of the ship itself' that Gardner had also followed in deciding the interior design of the *QE2*, as he recalled in 1988:

> The ship's walls are so interrupted by doorways, windows, etc, that I thought adding pictures and murals, usual on big ships, was a visual mistake. We aimed to make the structures themselves as elegant and easy on the eye as practical; real 'ships'.[38]

'Real ships', however, like the 'real world', are easier to cite than to define, not least because an element of caprice sometimes has its place in the shaping of a vessel, and especially in the crucially important styling of a liner. Riccesi writes of *Victoria*:

> She was agile and slim, and proposed new terms in the language of naval architecture, and at the same time suggested a point of mediation with the past through the recovery of certain elements of traditional shipbuilding displayed with self-conscious playfulness (windows, bridge front, railings).[39]

As we have seen, in recent years the spectacular growth of the cruise market has meant unparalleled new construction for many companies, and the high prestige (and cost) of cruising has encouraged a policy of innovative design; potential passengers need to see a striking vessel from the moment they open their glossy brochures even to think of parting with several thousand pounds for a luxury cruise. Joseph Farcus comments that 'a relatively small part of the traveling public has yet to take a cruise', and that such holidays are 'a very good secret which is slowly being discovered'. Robert Tillberg, too, notes that still only some five per cent of holidaymakers take vacations in the form of a cruise, but he also wonders whether construction has peaked. His own design company is looking to a future beyond the present cruise boom. This market is competitive as well as lucrative, and the effects on ship architecture have been considerable: 'The passenger ship is not dead: but to stay alive it has had to specialise and modify itself aesthetically and functionally, so adapting itself to this new dimension'.[40]

The need not only to accommodate large numbers of passengers in cabins but also to provide them with spacious public rooms and transit areas, internal and external, has meant that traditionally liners have had more extensive superstructure than other vessels. Only the car carriers of recent years exceed them in volume of enclosed space above the waterline. Liners have also always been vessels whose external architecture is remarkably symmetrical both longitudinally and transversally, and in this respect they differ from cargo vessels, whose specialised needs have in recent years produced some dramatically offset volumes and gear. Such solutions, even if required, would probably be considered unacceptable in a passenger liner whose overall style and symmetry is an

important aspect of its public appeal. In this respect the design of passenger vessels, despite some of its extravagancies of recent years, is arguably inherently conservative compared with that of cargo vessels.

The location within the hull and superstructure of large communal spaces such as saloons, dining rooms, swimming pools, grand staircases and atriums has always meant a need to accommodate windows as large as possible (once a calculated risk on an ocean-going vessel), together with additional overhead natural illumination through lightwells and skylights. The placing of windows in public rooms situated in prime areas such as the lower bridge front, a position distant from engine and propeller noise and with excellent views on three sides, also had aesthetic importance in punctuating a prominent facade. It was here that the larger windows – as opposed to portholes – of First Class Saloons were often to be found. Recent cruise liners have supplanted any such 'punctuation' with complete curved walls of glass whose design has as much to do with one-class accommodation for passengers coming from a different mix of society as it does with modern sophisticated techniques of continuous glazing and lightweight alloy construction. Where tall, elegant and curtained windows once marked out the First Class dining room and gave those within discrete yet private views, modern cruise ships invite all the carefree and public gregariousness of a leisure centre. An important exception to this principle is the increasing popularity of the private balcony on cruise vessels for those who can afford them as an addition to their cabin suite, a development which parallels hotel practice and replaces earlier open decks with private balconies giving access to sun and sea. The 'all outside cabin' (AOC) concept was developed by the Finnish Wartsila Shipyard (now Kvaerner Masa-Yards) in the early 1980s, and partially applied first to P&O's 1984 liner *Royal Princess*, on which 152 of the 600 outside cabins had verandahs. Later vessels developed the concept, which has been described as probably the 'most significant shipboard innovation since individual cabin plumbing'.[41]

In some respects sophisticated cabin plumbing and outside balconies have removed some of the excitement of ship travel, at least for incurable romantics; indeed, ship *travel* has itself become next to impossible. I sought to make the return leg of a 1974 visit to Australia by ship, but the only cabin available was a multiple below the waterline on an Italian emigrant liner, itself one of very few ships still making regular passages. Paul Fussell recalled a similar frustration four years later in attempting to travel the Pacific by ship:

> I saw myself lolling at the rail unshaven in a dirty white linen suit as the crummy little ship approached Bora Bora or Fiji in a damp heat which made one wonder whether death by yaws or dengue fever might be an attractive alternative. Too late for such daydreams. I found that just as I was enquiring, passenger ship travel in the Pacific disappeared, in April, 1978, to be precise.[42]

Cruise vessels are now the only passenger ships that call at most ports worldwide, although many cargo companies have found it commercially attractive to include passenger cabins on their vessels. On some, such as the fast banana carriers that operate between South Wales and the Caribbean, style is important and dressing for dinner obligatory. Fussell's linen suit survives only on the nostalgic silver screen, as ship design changes in recent decades mean that anything more than a brief archive shot of a period liner is impossible without costly special

*The 1992* Costa Allegra *(built in 1969 as the American container vessel* Annie Johnson*) showing stylish exploitation of the original transom stern (typical of container ships) by bold 'glass wall' stern superstructure*

effects; ships have ceased to be the romantically mobile settings they once were – perhaps because modern cruise ships lack the class-conscious ambience that was always part of earlier liners – and a golden era of cinema has closed. *The Poseidon Adventure* (1972) seems a suitable commercially successful film epitaph, and *Raise the Titanic!* (1980) an appropriately disastrous one.

The varied exterior surface, and the differentiated scale of architectural components made earlier liners more immediately comparable to the irregular and complex skyline of a city. Even today, most cities still offer such variety amid the recent highrise; in both Helsinki and Stockholm the modern Silja and Viking ferries which link the ports dominate the city skyline. In the great passenger liners of the interwar period a significant feature of their appearance was their elegantly curved hull and round-fronted superstructure, linking forms which were both punctuated and given dramatic expression by their funnels; this tradition found its final expression in P&O's *Canberra*. The effect of such rounded and stepped superstructure is well described by Riccesi in analysing how the design of the 1931 Italian transatlantic liner *Conte di Savoia* gave the ship a sense of urgency, and of speed. Riccesi characterises the vessel's bridge front as follows:

> Another detail significant in both architectural language and meaning was the front of the bridge: the gradual building up of volume in terraces, with the mass receding as the height increased, made the ship appear more slim and streamlined also in the superstructure, which seemed sculpted by the winds.[43]

From the liners of the nineteenth century right up to the *QE2* of 1969, whose design intelligently evoked but also reinterpreted the visual appearance of earlier vessels owned by Cunard, modulations of such architectural language conveyed the luxury, elegance and reliability of such vessels. It was a language appropriate to the times, comparable with some elements of land architecture, yet in its specific configuration peculiar to the well-understood aims and functions of a passenger ship. Such vessels tempted the potential passenger with glimpses of terraces – Riccesi's term is precisely right in its simultaneous architectural and landscaping associations – which offered views, exercise and seclusion alike.

Earlier liners were linked in many ways to the life and social patterns of the cities between which they plied, and were occasionally themselves depicted as one aspect of metropolitan life. Allan Sekula suggests a reading in these terms of the classic 1907 Alfred Stieglitz photograph, *The Steerage*, taken on board what Stieglitz himself described as the 'fashionable ship of the North German Lloyd', the 1902 *Kaiser Wilhelm II*, bound for the States:

> what Stieglitz discovers in *The Steerage* is a maritime space that becomes metropolitan, vertical and hierarchical, jam-packed with anonymity. The steamship, the largest machine capable of being pictured in its totality, yields a partial image of that larger, unpicturable machine: the city.[44]

The explicit ranking of a liner's terraced decks was once not only an essential part of the ship's formal composition as a structure but also of its function; passengers could leave port, or pass the time of day at sea leaning on the polished rail of their allotted deck, and never meet with those whose lifestyle, at sea as on land, was very different from their own in wealth, style and attitude. Sekula notes that in his 1924 study *Technics and Civilisation*, Lewis Mumford saw 'the big ocean steam-

*Superferries at Helsinki dominate the city background*

ship' as 'a diagrammatic picture of the paleotechnic class struggle',[45] and some Italian passenger ship design between the wars apparently reflected a similar pattern:

> The Ligurian school of naval architecture, even in the twenties, remained faithful to the English tradition of the ship as a floating symbol of class distinctions. There was a gulf between the furnishings of First Class accommodation and of that allocated to the poor – almost always emigrants or their relations, visiting from 'the distant America'; a gulf greater even than that between 1st and 3rd on the railways.[46]

Unlike Concorde, whose sleek shape not only facilitates but also expresses its supersonic speed, streamlining in ship superstructures, subject as they are only to wind resistance at relatively low speeds, serves generally to express those hidden developments in hull design and propulsion which can alone enhance a vessel's speed. The *Conte di Savoia* was fitted with a bulbous bow, a technology transfer from current warship design, but her modernity was proclaimed to the public by more visible features:

> there was now great insistence on the aerodynamic quality of the ship above the waterline, slimming funnels, rounding and stepping back deck levels (the *Conte di Savoia* almost resembled the terraces of a hillside cultivated in steps), raking masts and funnels, when the most significant results of the speed trials were basically to be credited to the power of the engines and to the underwater form of the hull.[47]

The history not only of liner design, but of critical commentary on it, illustrates some of the problems in discussing ships more generally in the context of architecture and design. The apparent similarity of most liners as compared to a range of cargo vessels has allowed critics to overlook differences in climate, market and company policy which affect basic elements of both exterior and interior design. The over-concentration, too, on transatlantic liners is explicable more in terms of fascination with those national rivalries which they themselves reflected than of any serious approach to the complex issue of liner design itself. Liners and passenger-cargo liners operating on no less important routes, such as those to India, Australasia or South America, have been largely ignored, as has the extensive network of routes which earlier served closer European colonies in the Caribbean and Africa.

An especially interesting vessel was the 1964 passenger-cargo vessel designed by Marshall Meek to carry both passengers and cattle between Fremantle and Singapore – suitably named *Centaur*. This tricky brief necessitated ventilators from the holds taken high up the masts but overall had a decisive sense of coherent design through sure balance of mass and detail.[48] Equally important in the very unusual and specific conditions of operation are those modern vessels now replacing the long established traditional ships on the 'Hurtigruten' services along the extensive coast of Norway, ships on which the constant proximity of land, the special lighting and weather conditions in such northern regions, and also the strong traditional consciousness of the Norwegian passengers who regularly use them, have produced very specific design and function characteristics.[49]

For seafarers themselves there has been something of a watershed between those vessels commonly trading until the 1970s, which in both cargo and passen-

*A difficult design brief which produced strikingly elegant lines, the 1964 Alfred Holt combined passenger and cattle carrier was appropriately named* Centaur

ger trades were characterised by 'distinct liveries as well as architectural styles',[50] and the modern generation of internationally designed and built bulk carriers and container ships. A motorman in his fifties commented in the late 1980s of the ships of Harrison's and Blue Funnel:

> These were our landmarks. They were part of the world. They were our world and they've gone.[51]

Such ships, whose builders and owners could be clearly identified from their colours and lines, became for the seafarer 'as much a part of the mental map of the landscape as a river or a mountain and with the same apparent quality of being fixed in time and space'.[52] They carried with them, in specific elements of their overall design and tradition, the long-established and seemingly unchangeable qualities of British shipbuilding and ownership. Through the continuity of their design, indeed a certain conservatism, the output of particular British yards and the distinctive vessels of particular companies overcame the transience of individual vessels to establish that sense at sea of familiar landmarks; the end, when it came, was all the more brutal.

In port a ship can present a large and unfamiliar presence in the local landscape, as Nöel Mostert found when his supertanker *Ardshiel* berthed at Pointe de Grave on the Medoc peninsular. Her size and the brevity of her visit made her something rather more than a 'figure' in the 'ground' of the rural landscape:

> For miles around, everybody's attention inevitably would be drawn to the gigantic ship, for she clearly had taken possession of a considerable portion of the estuary and its sky. Standing as a light frail vertical mark beside it, as though to provide a measure, the church steeple of Le Verdon was hardly noticeable.[53]

It is a mistake to see ships either as industrial design or as architecture; they are both, and only by making the effort to find appropriate terms in which to express this duality can they be fully understood:

> The ship is a machine for transport and for living in. We can analyse it both as a product of industrial design or as architecture: neither one approach nor the other, by itself, can offer a complete and relevant understanding.[54]

This central point has been little discussed, and one result is that there is no accepted terminology in which to discuss the aesthetics of commercial ship design:

> What is surprising today is the almost complete lack of a specific literature addressing the topic, which surveys and analyses it with a methodology in accord with the prevailing criteria of industrial design.[55]

One Italian architect who did understand the importance of ship design was Ernesto Rogers, cousin of Richard Rogers and a member of BBPR, one of the most influential architectural partnerships in Milan. At a conference on architecture and ship design held in Trieste in 1960 he commented:

> The ship must be understood as an object of industrial design, but in its entirety, not as the sum of separate objects designed with care and others added without discrimination or justification.[56]

Such an inclusive approach has its problems, for differing elements in ships demand different solutions; and Rogers, typically, was thinking mainly of passenger vessels. Riccesi suggests, however, that properly to understand the ship in

*OPPOSITE: Lifeboat on the Hutigruten ferry* Nordlys*;
ABOVE: Bow of the* Nordlys

its complex entirety ['in tutto il suo complesso'] is to cut across boundaries as much in modern shipbuilding, where design work has become increasingly compartmentalised, as in architectural practice and debate.

Certain basic differences between architecture ashore and afloat are clear: with the exception of warships and of some merchant vessels, most ships in the West operate in and for the private sector; there are no low-cost housing projects or public hospitals at sea, although many countries still operate national restrictions and subsidies. Again, with few exceptions, the design of a ship is more constrained by prevailing operational conditions and well-proven operating solutions than an architectural project on land, despite recent developments which include revolutionary 'open' container vessels in which removable covers and powerful pumps replace the main deck to allow faster turnaround times in port.

Quite apart from the instructive comparisons of design and function to be made between ship and land architecture, the role of shipping in global transport continues to have a profound effect on some of the most visible and controversial architectural developments in cities and their surroundings. Nostalgia for liners as part of a recent but forever-lost cityscape is common: we can all recall (or imagine) with pleasure those days, seen as typical of the thirties but common until the sixties, when 'giant steamship funnels still closed the view down side streets on Manhattan's Lower West Side and London's Rotherhithe', and we can also accept such vessels as 'an unmistakable reminder of the international trade which once underpinned their respective cities' economies'.[57] Yet subsequent developments in international maritime trade have helped determine the differing fates of city centre development sites such as Covent Garden in London and Les Halles in Paris:

> both places have evolved in almost identical fashion, not because of any shared architectural ideology, but because refrigeration and giant articulated trucks put paid to any need for city centre wholesale food markets.[58]

Beyond these markets, where land meets sea, the giant trucks are themselves fed by even larger seaborne transporters; and in this highly-automated process traditional dockers have vanished, 'displaced not by the long arm of Le Corbusier, not by the garden city of Ebenezer Howard, but by the shipping container'.[59]

Modern port developments foster nostalgia for a vanished labour intensive, and exploitative past by their foregounding of that automation and technology our consumer expectations demand, but our sensibilities reject:

> Harbors are now less havens . . . than accelerated turning-basins for super-tankers and container ships. The old harbor front, its links to a common culture shattered by unemployment, is reclaimed for a bourgeois reverie on the mercantilist past . . . Everyone wants a glimpse of the sea.[60]

The sea we all want to glimpse, however, is now either deserted by trade (the 'old harbor front') or is the purely recreational one of yacht marinas; at the former Yorkshire fishing port of Hull, a shopping complex has been built on stilts in the centre of the former fish dock. On the other hand, modern shipping too openly proclaims its industrial nature to appeal to the general imagination, and in any case many of its processes are too dangerous to allow access:

> Factories become mobile, ship-like, as ships become increasingly indistinguishable from trucks and trains, and seaways lose their difference

with highways . . . This historical change reverses the 'classical' relationship between the fixity of the land and the fluidity of the sea.[61]

In global economic terms passenger ships are not, and never have been, the most important vessels, despite their high profile both in the public imagination and in traditional design commentary. As Mostert wrote in 1974:

> Oil tankers, a once obscure and largely unremarked race of ships, have established themselves during the past fifteen years as the dominant vessels of the age and, arguably, of all time . . . they are quite simply the principal means of carrying the world's oil from where it is to where it's needed. Without them, much of the world would simply stop . . .[62]

For their part, container ships are more important to the economy of London or New York or Paris than transatlantic liners ever were, but liners had the glamour and, in New York at least, berths against an incomparable skyscraper background. But as the rapidity of such liners' demise showed, their prestige was partly the visible expression of political as well as economic interests. In building and maintaining them, governments were doing at sea what they had often done on land, using architecture to impress: 'the towers of San Gimignano were built to symbolise the power of their owners as much as for practical purposes'.[63] It is a measure, however, of the extent to which the transatlantic liner has lost its once magical resonance – a loss symbolised by the ignominious survival of the former Cunarder *Queen Mary* at Long Beach – that Houston can now be compared to Los Angeles in the 1950s, 'before the bungalows in their orange groves began to be swamped by giant condominiums popping up in their neighbourhoods like beached Queen Marys'.[64]

*The 1951 Everard coaster* Seniority *leaves Brixham after bunkering, August 1960. These vessels had the 'three island' layout which would soon be replaced on most vessels by the 'all aft' configuration*

# FUNCTIONAL ELEMENTS OF
# SHIP ARCHITECTURE

*Nobody can draw ships until they have examined them closely. You must know what are the essential parts of a ship; what can be left out and what must be put into a drawing. A ship is like the body of a human being, with bones, muscles and sinews. The 'clothes' which cover the body of a ship are the least important parts.*

Peter Anson, How to Draw Ships, *1955*

An understanding of how ships are built and operated has often been prevented by ill-informed or careless comments on the part of architects and critics lacking the necessary interest or insight. The best way to understand any machine with moving parts is to watch it work; lack of any moving parts is at once the miracle of computers and the reason why they are so profoundly boring as objects. For its crew a ship is quite literally a machine for living, but for its owners it is a mobile machine for earning a profit, and much of this chapter will look in some detail at the working elements of modern ships. Any comprehensive survey of the range of vessels operating today would require a larger, and different, book. Instead, some examples of ship design are here used to illustrate more general points, with three sections dealing with key elements of ship design: the hull, the super-structure, and rigging and gear. As we shall see, developments in the design of some modern vessels have made such a tripartite division meaningless, but it does reflect traditional patterns of building and function and so may be taken as basis against which to measure recent departures.

## I Shipbuilding

Ships have traditionally been constructed on the shore not only for ease of launch but because water was the original means of transport for many of the materials required in the process. In many countries sizeable wooden fishing vessels are still built thus, using traditional designs adapted to motor power:

> Ships were built on age-old principles which changed little from century to century or place to place. Through the years the design of each part and the particular cut of the wood had been dictated by generations of men observing ships in action in the wild element of the sea. From this accumulated wisdom evolved shapes of timber which were efficient and pleasing.[1]

The hull of a vessel was usually assembled from the keel up, piece by piece, in the open air and was finally launched down a slipway into the water. An alternative method for smaller craft is the 'shell first' technique, in which the stiffening ribs are inserted after the hull form has been completed. Early modern shipyards industr-ialised vernacular skills, although the stresses imposed by propulsion machinery demanded new solutions, and especially with larger iron and steel vessels the

OPPOSITE: A modern cargo vessel, the Frederikshavn Denmark *under construction; ABOVE: Launch of the* Aquitania *at John Brown's, Clydebank, 20 April 1913*

*FROM ABOVE: The 1995 Hammann & Prahm shortsea trader* Rebecca Hammann *on the slips before broadside launch, Haren; Bow view of the* Rebecca Hammann *directly after launching; Stern view of the* Rebecca Hammann *directly after launching*

need for metalworking facilities went alongside the advantages of rail links, which grew alongside the new technology.[2] Smaller fishing and other powered vessels were built by hand with traditional methods well into this century, and one shipbuilder who was apprenticed at Lowestoft in 1916, and worked in the same yard for almost seven years, has produced a fine illustrated record of the process of building a wooden steam drifter, 'from tree to sea'. Such methods survive today mostly in yards building traditional yachts. Shipbuilding has, like other high technology industries, come to rely increasingly on buying in ready-made units from skilled and specialised suppliers.[3] It was the launch, accompanied by traditional ceremonies of naming and blessing the ship, which marked the start of her seagoing life; the 'topping out' of a building seems a poor substitute, as does the modern preconstruction of large elements of a hull prior to their assembly and eventual 'floating out' as a near complete hull from a covered yard. An alternative method for smaller vessels is a sideways launch, which is possible even into quite a narrow waterway since the vessel stops more quickly, even if launched at speed; Dutch shipbuilders have long used this method for the construction of small cargo vessels. Illustrations here show the 1995 German shortsea trader *Rebecca Hammann* moved down to the water's edge ready for a more gentle sideways 'launch' into the River Ems against the idyllic background of the town of Haren. Clearly visible in the bow view after launching is the grille covering the bow propeller which provides transverse thrust for docking and manoeuvring in confined spaces; a stern view shows how both internal hull and deck space are maximised by the vessel's transom stern, which also includes a stern anchor (useful when mooring in crowded waterways) recessed to leave clean hull lines for docking. Also just visible is the circular shroud ahead of the propeller designed to direct water flow and increase efficiency.

Such small modern vessels are often designed to high and precise specifications, with a view to the configurations of specific ports, although this in no way prevents them operating further afield, as a report on this particular series of vessels for *Lloyd's List* outlined:

> Like the series-opener *Heyo Prahm*, the new shortsea trader has been optimised for the traffic in packaged timber to Beckingham, opposite Gainsborough, on the River Trent. The tight bend on the tideway at Morton Bank, just downstream of Gainsborough, was the main determinant for the design's fifty-eight metre overall length.[4]

David Tinsley went on to note that, while 'optimised for the Gainsborough trade', this class of vessel:

> lends itself to operation into a range of small ports and berths on broad-gauge European waterways, including the Rhine, (to Duisburg), the Seine, and the Albert Canal. It is also suitable for trade on the Saimaa system in Finland.[5]

In these sea and river trades small vessels maintain that intimate link with local towns, through both their construction and cargoes, which has been mostly lost in the deepsea trades. Indeed, the vessels specifically designed and built for importing timber from Sweden to the Trent have prevented the decline of local trade:

> by introducing a class of vessel with virtually double the timber intake capacity and twice the cargo handling productivity of the coasters previously used,

new economies have been brought to the traffic – and thereby ensured its preservation.[6]

The illustration of the *Lore Prahm* outward bound on the River Trent well illustrates the flexibility of such vessels.

In architectural terms the period during which a large vessel is under construction ashore makes possible comparisons between the steel skeletons used in many modern buildings since early developments in Chicago in the 1870s and those which form the framing of a ship, although these are designed to withstand greater strains than any static building will experience. Whatever the process employed, one of the fascinations of a shipyard is that seemingly chaotic process by which myriad parts and fittings, whether made on site or bought in from specialist suppliers, fit together in a carefully planned sequence to form one strongly unified, seaworthy and independent shape. There are obvious analogies with a building site, but always with the difference that this structure is designed to sail away. It is during this building period, too, that comparison is possible between the scale of the growing vessel and the surrounding factories and houses, though modern shipyards are increasingly located away from residential areas and are also covered.[7] If built in the open air, the land-based period of a ship's life is a dramatic demonstration of its internal intricacy and integral strength; Mostert was also struck by the immense scale of his supertanker on descending into the cargo tanks at sea for cleaning and checking, and drew imaginative analogies with both land and sea architecture:

> I hadn't expected quite such a mass of detail, of elaborate patterns of pipes and valves, so many flights of other ladders, so many beams and struts, platforms, nooks, crannies, and crevices. Most impressive of all were the transverse beams thrusting out into the huge emptiness, like transepts.
>
> Returning up the ladders to the main deck, I was again struck by that anomaly of these ships whereby they retrieve in so many curious ways some of the experience of the great days of sail. The ascent and descent of great heights is a recurrent part of ship's routine aboard supertankers to a degree that hasn't been known since sail . . .[8]

A traditional launch is itself a very considerable strain on the ship's essential structure as within a few minutes several thousand tons of metal move from dry land to water. James Steele's book on the *Queen Mary* notes that the liner flexed on the vertical plane almost eight inches one way and some three inches the other during launch, within normal tolerances and expectations but an indication of the strains involved – many of which a ship is designed to survive also in service.[9]

After the launch, the lengthy and laborious process of 'fitting out' traditionally began, with the vessel now moored alongside the shipyard quay. Much of this fitting out process is now completed in a covered yard, with the largely completed vessel gently floated out without the risks associated with traditional launching methods. The whole process of fitting out is very labour intensive and, even for a cargo vessel, demands extensive skills in electronics, metalwork, plumbing and engineering, with the quality of the finished ship largely dependent upon the dedication and reliability of each individual in a large workforce. The wide range and high quality of skills required for passenger vessels has meant that relatively few have been built by yards in the Far East, compared with the success

*The 1989* Lore Prahm, *a sistership of the* Walter Hammann, *heads seawards on the River Trent*

of such yards in building comparatively simple vessels such as tankers.

In contemporary photographs the hulls of liners built between the wars, a period which included the two 'Queens' for Cunard, tower over the houses of those who built them, and such architectural incongruities partly explain why such vessels occupied a special place in the affections of the local workforce and the nation alike. In scale these great vessels were indeed transient cathedrals and were perforce built in flat areas where, once launched, the immediate dockland landscape reverted to its predominantly drab, flat profile once again. No land-based structure has ever evoked such affection in Britain as did these ships. While shipbuilding and its associated industries provided substantial employment, and also helped maintain the nation's imperial role, the public's affection for these vessels was complex, certainly questionable, but of undoubted power:

> It required an age of romantic public involvement with, and affection for, the whole body of shipping, complementary to mere pride and patriotism, to bring the view of ships and the feeling for them that we have retained until today. This required a revolution in shipping itself, which the early decades of the nineteenth century provided. Modern shipping and modern seafaring as we have known it as well as our feelings about it are mainly a nineteenth-century creation, and substantially British in origin.[10]

The modern parallel to such economic and social importance of shipbuilding is perhaps to be found in the all-powerful Hyundai Corporation in South Korea, which builds everything from telephones to cars to supertankers; their current press advertisement stresses that their industrial output ranges 'from chips to ships'. Allan Sekula has speculated on how, in the modern age, the Corporation's dominant role in social as well as shipbuilding engineering might determine the reactions even of those seeking to place contracts:

> In Ulsan, once a mere fishing village on Mipo Bay, it is possible to stay in a Hyundai-owned hotel, schedule appointments by Hyundai cellular telephone, shop for a bottle of Scotch whiskey or soju or a packaged snack of dried squid at the Hyundai department store, and travel in a Hyundai automobile – manufactured just down the street – to meetings with the officials of the world's largest shipyard, a division of Hyundai Heavy Industries. The situation provokes paranoid speculation.[11]

The last fifty years have seen greater changes in shipbuilding techniques than at any time previously, and many derive directly from wartime loss replacement programmes which demanded fast construction methods suited to a relatively unskilled workforce, many of them 'Rosie the Rivetter' and her invaluable sisters. Standardised and simplified designs, together with largely welded construction and the offsite pre-assembly of some units for the purpose-built 'Liberty' yards, brought the record delivery time for one US-built 'Liberty' ship of less than five days from keel to launch (compared with around five months for early ships in the programme). These vessels finally numbered some 2,700, out of which 700 were still trading twenty years later. They dominated the postwar freight scene, and the innovations they represented had a lasting effect on the shipping market and on subsequent shipbuilding practices, not least in successful Liberty replacement vessels such as the British type SD14 (Shelter Deck, 14,000 tons deadweight) cargo vessel constructed from 1967.[12]

*OPPOSITE, FROM ABOVE: A photograph taken in about 1868 of a Helsinki shipyard; today it is the Helsinki site of Kvaerner Masa-Yards; The same site from the air in May 1929. Kvaerner Masa's covered building yard is now on this site – see page 84*

*The same site in October 1993, with two liners for Carnival
Cruises fitting out*

The trend towards the separate assembly and finishing of increasingly large elements of a vessel, and towards overall construction of the whole vessel in a covered yard with controlled conditions, is illustrated by developments at the Kvaerner Masa-Yard in Helsinki, a major shipbuilder whose modern covered dockyards still occupy a site near the centre of the city: when I was there in 1995 an almost-completed liner towered over the quayside outside my hotel. A photograph of the original drydock site in 1868, today intact but unused, illustrates the difficulty of work in such open conditions, while comparison of the site in 1929 with its present facilities shows how covered building yards have replaced open air sites. Final fitting out still takes place with vessels moored alongside the quays, and an October 1993 photograph (from the top of my hotel) shows the two Carnival Cruise liners *Sensation* and *Fascination* nearing completion.

A local pioneer in building ships under covered conditions since 1970 is Appledore Shipbuilders, a small British yard in North Devon, several of whose ships have already been illustrated. Their innovation was replicated by other and larger yards, and it was Meyer Werft at Papenburg, bidding with the close support of the local region, who were builders of the new 67,000-ton passenger vessel *Oriana*, launched in 1995 for P&O Cruises. Such 'factory' conditions for shipbuilding sharpen comparisons with, for example, the prefabricated washrooms for Foster's Hongkong and Shanghai Bank – these were assembled in Japan, to be fitted complete onto the building during construction – and developments in the fitting out of modern liners. Accommodation and bathroom 'wet units' are pre-assembled by Kvaerner Masa-Yards on just this basis for onboard installation as finished elements which can be removed in the same way at a later date; in this instance the hull decks retain their primary function as the 'foundations' upon which and within which accommodation is assembled, albeit with different technology; cabin and washroom units are complete but the floor is the deck itself, thus facilitating possible rearrangement in the future. Such modular construction is a far cry from the intricate plumbing, wiring and panelling of earlier passenger vessels which, like the office buildings and hotels of their era, were designed magnificently but inflexibly for a different, and seemingly more certain, age.

One aspect of shipbuilding only possible because of modern materials and construction methods is that of 'jumboising' a vessel by insertion of a new section, usually in response to changed market conditions or the need to gain greater efficiency in operation. In this process the new section is first constructed and the existing vessel then sliced through, usually amidships where lines are simpler, and the new section inserted. Occasionally a new bow section may be constructed, but this process is usually restricted to damage repair, or occasionally to essential alterations needed for better handling.[13]

Bulk cargo vessels are the easiest ships to jumboise, but passenger liners have also had their hulls and extensive superstructures lengthened. The techniques of jumboising owe much to emergency repairs during the Second World War, although shipyards had always built replacement sections for damaged vessels; it was the combination of lower labour costs achieved through prefabrication and welding that enabled the concept to be feasible in high-wage European yards. A more recent development has been the construction of vessels with a view to their being jumboised at a later date should changes require it.

*FROM ABOVE: The simple funnel and ventilators of the Liberty ship Jeremiah O'Brien, built between 6 May and 30 June 1943 and now preserved in full working order at San Francisco. In 1992 she crossed the Atlantic to attend D-Day events; The Liberty ship Ignace Pederewski, built Los Angeles 1943, at Southampton in December that year. More than 2,700 of these vessels were built in the USA between 1942 and 1945*

## II The Hull

The hull of a vessel is essentially a strong container formed of a relatively thin metal skin supported by a frame of internal girders, but whereas the skin of many modern buildings is only decorative, that of a ship's hull is vital for safety in ensuring the seaworthiness of the entire vessel. Even the most basic 'ship', such as a Thames lighter, requires this, and on many larger vessels the double-bottomed hull has great intrinsic strength and overall resembles a box girder more than a mere container. At the stage of fitting out large openings may be left for access to the interior, with the result that the hull hardly resembles a ship at all. Such openings will be welded up as fitting out is completed, and when a ship is made ready for sea all hull openings must be regarded as potentially dangerous and provided with secure methods of closure, just as on an aircraft; only rust streaks from salt water indicate what such fixtures must withstand. Attention is also paid to ensuring a smooth hull surface which will not snag cables during docking and will allow tugs alongside when needed. Window openings in older craft, or in those operating in sheltered waters, are positively domestic in scale but on oceangoing vessels they are of thick glass and inherently strong in shape, and if close to the waterline may be portholes, possibly also fitted with a strong metal flap which can be screwed down from inside the cabin. The larger the opening, or the more subject to battering from the waves, the more crucial that it be secured when the vessel puts to sea.

This need was tragically demonstrated by the foundering of the 1980 passenger and vehicle ferry *Estonia* in 1994, probably through failure of her bow door securing mechanism, with the loss of over 900 lives. In port passengers can access any vessel well above the waterline by elevator and covered walkway but only the large openings that ferries have at low level in their hull structure make possible the rapid turnaround of heavy motorised cargoes and passenger vehicles, as well as train wagons. Some vessels have access from bow and stern, enabling vehicles to drive straight through, and a variety of systems are employed to open and close the necessary hull openings, from complete lifting bow sections (with further watertight doors inside) to divided doors which fold back flat against the hull sides, and are therefore a constraint on its profile here. The deck plan of such ships makes clear the extent to which, at rest, they function as an extension of the land transport network and docksides have specially designed terminals at which ferries fit snugly to the landmass; other vessels carry their own immense loading ramps with them, and some can swing these at an angle to load and unload from any quay alongside. The primary function of the ship as floating container is immediately apparent in such ships, just as bulk carriers dramatically illustrate the extent to which a ship functions not only as storage space but also as a workplace in its own right. At sea, though, it is a different story and the *Estonia* disaster, like the earlier capsizing of the *Herald of Free Enterprise* off Zeebrugge, has brought calls for the full-width decks on ro-ro (roll on-roll off) ferries to be fitted with watertight barriers to prevent any water shipped swilling from side to side, with rapid loss of stability.

Any hull possesses in differing degrees design features which enhance its ability to withstand the demands of the sea. A hull must penetrate the water at some speed, and at the waterline the 'active' and 'passive' elements of overall

*FROM ABOVE: This Thames lighter illustrates the very basics of cargo shipping; The Carnival Cruises liner* Inspiration *fitting out at Helsinki, 1995*

*FROM ABOVE, L to R: Ferry door, Southampton; Rolling steel hatchcovers on the German bulk carrier* Hera, *Tilbury; Fire-fighting gantry, Red Funnel tug, Southampton; Nautical detail on the 1955* SS Shieldhall, *now preserved in working order at Southampton*

*The ferry* Duc de Normandie

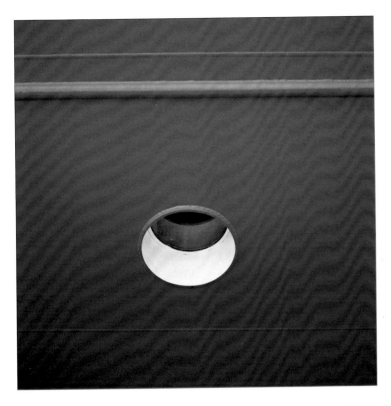

*FROM ABOVE, L to R: Porthole on a cross-Channel ferry;
Bow doors, ferry; Windows on the 1993 Holland America
cruise liner Maasdam; Bow doors, ferry Pride of Bilbao*

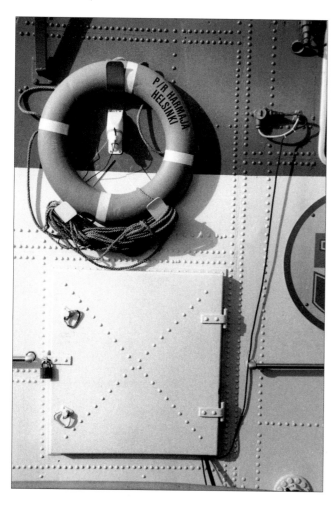

*FROM ABOVE, L to R: The delights of well-designed detail – superstructure riveting on a Swedish rescue vessel at Helsingborg, 1995; Nautical detail – a working steamship at Stockholm, 1995; Bow doors and ramp on the ferry* Duc de Normandie; *The ro-ro ferry* Gabriele Wehr *docked at Southampton*

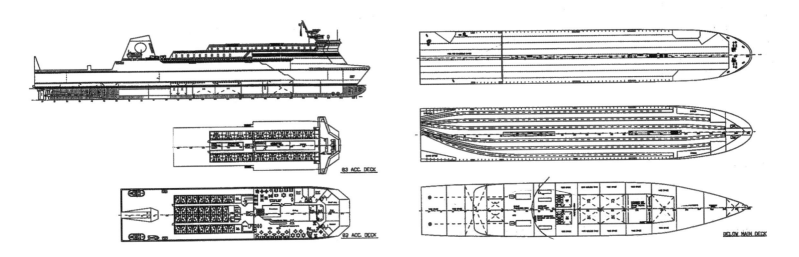

*FROM ABOVE: The ro-ro cargo vessel* Symphorine *at sea; designed by Skipskonsulent AS; General layout of a rail-truck-passenger ferry – plan by Skipskonsulent AS*

*The container ship Newport Bay straddled by cargo gantries*
*at Tilbury*

ship design clearly illustrate the differing demands that each makes upon the designer. For example, the necessarily square-sectioned hull and extended deck area of a container ship fines down to elegance at the waterline, below which the dynamics of water penetration and the provision of efficient buoyancy dictate a bulbous bow, which is completely submerged when the vessel is fully loaded. At the stern of the vessel water penetration is not required (though a dynamically efficient underwater form still is) and containers can be fitted square to the form of the transom stern.

The lines of such a cargo ship can be no less elegant than those of a passenger liner (though rarely do they have any decoration) for speed may be essential; other priorities produce very different shapes. Harbour tugs are designed to assist in the docking manoeuvres of larger vessels, and the inward-sloping bulwarks of their hulls, together with superstructure set well inboard, enables them to push directly with the heavy rubber fenders on their bows against the possibly flared sides of ships many times their size. Power, compactness and manoeuvrability are priorities for tugs, with every detail designed for speed and effectiveness in handling ropes and hawsers in what can be extremely busy and demanding harbour situations. The tug bridge must give excellent all-round visibility for her captain, the towhook must be easily accessible and the rear deck must be clear of all obstructions that might snag the tow and capsize the tug; even access doors are neatly inset to ensure uninterrupted deckspace for crew on a small vessel.

Specialised service and rescue craft demonstrate very clearly the extent to which hull form is shaped by function: *Cam Sentinel* has a high freeboard towards the bow, complete with helicopter pad, in order to operate safely in rough seas, excellent all-round visibility from her bridge (with the funnel eliminated) but low freeboard in her rescue zone to enable accident victims to be brought aboard. Even the vessel's colour is functional, ensuring maximum visibility in poor weather conditions, a practice now adopted by many modern vessels, and especially by tankers. The modern tanker *Dutch Engineer* is a fine example of clean proportions by the same Norwegian design firm Skipskonsulent, and has each functional mass of the vessel (hull, accommodation, funnel) sharply defined both by profile and paint; the visible codes of shape and colour here proclaim that security and efficiency achieved largely by what is hidden from view (hullform, engine choice, onboard computers and electronics). The *Dutch Engineer's* black funnel contrasts logically with her red hull and white superstructure, which suggests elegance and efficiency. The determined vertical of the square funnel mass gives scale and precision to the design; such well-balanced vessels are a pleasure to see despite the lack of traditional sheer curves.

### III Superstructure

The size, fame and status of ocean passenger vessels have always drawn the attention even of those who knew little of ships, from politicians to designers. These are the ships that the layperson is most likely to know by sight, if only from advertisements, and their prominent superstructure houses those grand public rooms beloved of academic critics as much as they were, in an earlier era, adopted by travel writers.[14] This is also the aspect of such vessels that has mostly occu-

*FROM ABOVE: Status symbol – bow decoration on the 1961 P&O liner* Canberra, *here at Southampton, continues a long company tradition; Contrasts in scale and shape – the Alexandra Towing Company's 1928 steam tug* Sloyne *passes the* Queen Mary *in Southampton's Ocean Dock*

*FROM ABOVE, L to R: The Lykes Lines container vessel*
*Tyson Lykes and tug, Tilbury; Building for the sea – bulbous*
*bow and flare on the container vessel Newport Bay, Tilbury;*
*Bulk handling, but with manual assistance – Russian grain*
*vessel, Tilbury 1995; The ship as workplace – lowering an*
*excavator into the hold of Russian grain vessel, Tilbury, to*
*direct grain towards the suction gantry*

*Red Funnel tugs at Southampton*

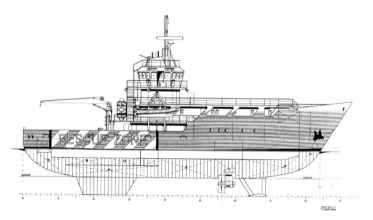

*FROM ABOVE, L to R: Tug at Hong Kong. Much cargo is still offloaded there from ships at anchor by large lighters, which are then towed by tugs to the docks for discharge; Pusher tug Miss Tammy on the Mississippi, New Orleans, 1994. Very large units of barges securely lashed together are propelled by tugs on the river. Miss Tammy is comparatively small; The shortsea tanker Dutch Engineer, designed by Skipskonsulent AS; The rescue vessel Cam Sentinel, designed by Skipskonsulent AS of Bergen, in the oil exploration context she was designed to serve; A full profile of the Cam Sentinel shows the sophisticated equipment to give both speed and manoeuvrability.*

pied the attention of architects commissioned for design work on board ships; the company can justify their fees because passengers will be suitably impressed with the sumptuous accommodation.

On any ship the superstructure is that part which can be treated, at least superficially, as a series of seaborne living spaces in isolation from the remainder of the vessel, especially if visualised from within; the ship's deck and the sea beyond are only glimpsed through the window, the ship's engines are several decks below. On a liner such a perspective is also that of the passengers, and therefore one that successful ship design must attend to, but as an overall approach to ship design it has serious failings. It is unhelpful for understanding the overall design of cargo vessels, which have a small crew and carry at most between ten and twenty passengers (most carry none); with such vessels superstructure is little more than a necessary inconvenience as far as cargo-carrying is concerned. In all ships, however, their total sense of proportion and design is determined very largely by the relationship between the hull and the superstructure, these being the two prominent masses in a ship's composition and theirs a relationship usually emphasised by form, contrasting colour values and paintlines.

This element of ship architecture is illustrated with classic elegance by a vessel shortly to start a new career, the British Royal Yacht *Britannia*. She is fittingly restrained and traditional in design, her decks receding like many a liner of the 1930s (although she was built in 1953) and their gently rounded front sections echoed by the understated softening of verticals in the superstructure, as well as by her large and gently-raked funnel. The traditional combination of red boottopping, dark blue hull, white superstructure and buff funnel serves her well, and is discreetly enhanced by a thin line of twenty-four carat gold leaf along her hull; with her boarding steps swung out the counterpoint of hull and superstructure is enriched further. A more distant view of *Britannia* against London's city skyline reveals the simplicity of her profile, and the importance of the large buff funnel in providing a central focus for the whole composition. *Britannia* is unique not only in the time and money spent in her upkeep (some £10 million each year) but in the measures taken in building her to ensure the highest possible standards and finish. It was feared that the slight distortions produced by the heat of welding would disrupt the mirror-smooth finish required on the hull and so it was riveted, each rivet head then being ground down.[15]

Passenger liners present different problems of scale and of finance. Sir Hugh Casson recalls walking the steel deckplates of *Canberra* when the ship was being built at Harland and Wolff's Belfast yard in 1960. As he pondered the possible design of a lounge, plans in hand, a small man in a cloth cap came up and marked the steel plates with a large chalk cross. Casson asked what this signified, and the man replied that it was the location of a vertical ventilator shaft from the engine rooms. 'It's not on the plan', observed a surprised Casson, to which the man rejoined: 'It may not be on the plan, but it's on the floor now'.

The moral of the incident, as Casson wryly recalls, is that with liners the designer of passenger space is always at the mercy of the naval architects, engineers and safety inspectors who fix the location of essential structural members, watertight bulkheads, fire walls, ventilator shafts and the myriad other elements of the vessel's labyrinthine structure. That unmarked ventilator shaft would have to

*FROM ABOVE: The 1953 Royal Yacht* Britannia *with boarding steps lowered; Detail of recessed door, Red Funnel tug Southampton*

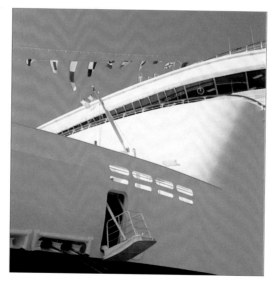

*FROM ABOVE, L to R:* Canberra's twin funnels proved a design landmark; *In shaping the funnel for P&O's 1995* Oriana Tillberg Design *sought to evoke* Canberra's *twin stacks in a modern profile; Wooden decking on* Oriana; Oriana *and dockside architecture; High tech –* Oriana *superstructure front; Fine design and good maintenance – Bridge front on the 1993 Holland America cruise liner* Maasdam

*FROM ABOVE, L to R: Modern lines and traditional wood – rear decks on Oriana; Familiar square format – Oriana from the stern; Deck view on Oriana; Nautical perspective – steel decking on Oriana; Attractive harmony of contrasting materials for specific purposes – steel door and wooden deck, Oriana; Practical above all else: tug funnels, Southampton*

be accommodated within the lounge layout, possibly by a dummy 'column' elsewhere in the room for symmetry. The vessel must also have external lines which suggest the elegance within, and here *Canberra* is an interesting case. Because she was built as long ago as 1961 her hull still has basic sheer along its length, and this is subtly accentuated by the flowing and sculpted quality of her superstructure, following and developing as it does these curved forms with great elegance and simplicity. As we have seen, P&O's new *Oriana* is intended deliberately to evoke the style of *Canberra*, not least through the 'twin' moulding shape of her funnel, but the comparative squareness of the new ship shows in comparison, despite sweeping deck shapes at the stern and the generous use of wood to soften the onboard ambiance. *Oriana*, like her modern competitors, has what Gardner considered an economy tailoring in her overall shape, and where *Canberra's* sweeping lines contrast sharply with square-edged dockside machinery, her more modern sister shares a certain affinity with it. This is not to say that one design is better than the other; they reflect the difference of their respective times.

Well above sea level the designer can today allow those large openings in the superstructure which always define important public rooms, and indicate even to those ashore the elegant facilities the ship offers its passengers. On occasion such elegance can owe as much to simplicity of design as to decoration; from clear expanses of deck to the spare forms of welded superstructure and the specifics of doors and fittings, a ship has much to offer in terms both of architectonic mass and intricate yet practical detail.

In passenger liners superstructure played an important role in form and function from the early days of steam, but many cargo ships had only minimal open superstructure well into the second decade of this century. Crew accommodation was situated in the hull itself at bow or stern and only officers' accommodation and navigational requirements shaped the superstructure. In recent years the combination of vast size, high automation and (consequently) low crew numbers has meant that modern bulk carriers have returned to a pattern of minimal superstructure, with their bridge sometimes raised on narrow columns in order to give the necessary height for navigation; the bridge of the Japanese coal carrier pictured here in Rotterdam's Europort towers above her discharged cargo is like a building in a self-constructed landscape. Although there were earlier examples of tankers with minimal bridge structures, such designs became increasingly common as ship size grew rapidly in the 1950s and 1960s and modern vessels can offer designs of striking clarity and proportion.[16]

The origins of the ship's bridge[17] itself as a protected steering position may stretch as far back as the late sixteenth century, but more recognisable origins lie in the era of paddlewheel propulsion when, in the 1870s, it was found convenient to 'bridge' the two amidships paddle boxes with an elevated walkway across the hull which gave a good view for navigation of the vessel. Prior to this development many ships were steered just as sailing vessels had been, from the poop deck located aft. Officers had their accommodation here, or amidships, with the minimal crew accommodation remaining in the fo'c's'le. Steering arrangements on the 'bridge' were completely without protection from the weather, and even when this moved from aft, weather protection was often only canvas screens which deflected wind and rain; such vessels thus retained many of the appalling work-

*FROM ABOVE: Dramatically curved bridge front on the 1961 P&O cruise liner* Canberra *emphasises her flowing lines; The* Britannia *against the city of London skyline; OPPOSITE, FROM ABOVE: Landscape with bridge – the Japanese coal carrier* Ijmuiden Maru *at Rotterdam's Europort, 1986; Bulk carriers in Rotterdam's Europort*

*FROM ABOVE: Architecture in the making – the 1892 paddle steamer Koh-inoor, which ran between London Bridge, Southend and Clacton, is seen here on 13 August 1906. The open 'bridge' between her paddle boxes is clearly visible; An unidentified paddle steamer in the Channel, also in about 1906 – her bridge and wheelhouse has a slightly different configuration; The most elegant and evocative of partings – in 1948 the P&O liner Orcades, built that year, pulls gently away from the quay at Sydney*

ing conditions from the supposedly glorious days of sail. As on sailing vessels, the galley (kitchen) was located on deck, partly for safety but also because in good weather there was the chance of the cook working with an open galley door (which is often of divided 'stable' design on small vessels even today).

The form of a vessel above main deck level was originally determined largely by the demands of human accommodation, whether crew or passengers; in modern cargo vessels clear access for dockside machinery to cargo space is a prime consideration, and this has produced vessels on which the profile of crew accommodation is increasingly 'high rise' in order to occupy as little deck space as possible. Some new container vessels have almost eighty per cent of their deck space free for containers, and even bulk cargo vessels are now often built with provision for carriage of containers on their steel hatchcovers, and the wheelhouse must also offer clear sightlines over such possible deck cargo.

Until the late 1950s and the worldwide impact of containerisation the relationship of hull to superstructure was reflected clearly in the construction, paintwork and paintlines of a vessel: varnished wooden doors and panelling, white-painted metalwork, scrubbed wooden decks and brass or braided rope handrails for the bridge area; polished brass window and door fittings; painted steel plates for the hull. *Britannia* is a uniquely well preserved example from this period but at the other extreme, the 1955 *Shieldhall*, steam-driven like *Britannia* but with an altogether more prosaic function (the disposal at sea of sewage sludge), provides an excellent example of ship architecture in its most basic yet elegant form. Her bridge front has clean and regular proportions, with the shape and dimension of the enclosed wheelhouse windows echoed throughout its whitepainted front, while the wooden decks (unthinkable now on any working vessel) are tidily arranged. Everything here is functional, nothing for show, but the effect of thoroughgoing practical coherence is enhanced by the consistent adoption of traditional colours for machinery and detail throughout the ship.

The paint colours used on most ships up to the period of containerisation in fact derived from a much earlier and entirely functional palette from the era of sail: wooden hulls were then tarred black to preserve them, while wooden deckhouses were painted white or varnished; the buff colour still mostly favoured today for masts and cargo derricks derives from the varnished wooden masts and spars of sailing vessels. Brown or 'varnish effect' was widely used for steel wall plating on many small steam cargo vessels; with coal firing (and with coal often as a bulk cargo) it made sense to have colours which did not show every mark, but the effect was decidedly dark and drab compared with many bright and clean vessels today. The demise of breakbulk cargo and an increased emphasis upon shipping as one element in a highly mechanised cargo-handling system operated by consortia of international companies brought adoption of bright colours – red, blue, orange, green – comparable to those of heavy earthmoving machines, industrial plants and of some architectural structures and detailing ashore. The Italian Contship line has a policy of painting its vessels in a variety of all-over colours never before used at sea (such as shocking pink), which necessitated special research by the supplying marine paint companies. Consciously or not, such colours emphasise the role of these vessels as post-modern 'engineering' structures, as machines, rather than evoking a visual language which looks to the days of sail or steam; a parallel land development can be seen in the use of bright colours (especially blue, green, yellow and red) for metal detailing on many contemporary buildings.

The increasing scale and complexity of some shipborne equipment, such as vehicle discharge ramps, also means that these form a major part of the vessel's overall profile; traditional cargo derricks were at least reminiscent in form and scale of the spars and booms on earlier sailing vessels, whereas modern gantries and cargo discharge equipment are very visibly machinery that goes to sea. Ships have always been exactly that, of course, but the language and colour of modern design makes this very apparent, and nowhere more simply than in that common element of much transport and storage today, the ubiquitous container.

The distinction between hull and superstructure was clearly marked on most passenger vessels until the 1960s by the contrast between the closed structure of the hull and the relatively open, multi-deck arrangement of the superstructure. Where the body of the hull presented a solid surface pierced at most by portholes, the superstructure was characterised by a series of doors, stairways, windows and portholes at the various deck levels; this was true of both cargo and passenger vessels, but in the latter open decks also provided communal space (within ticket class restrictions) with even the exterior profile of such vessels reflecting this. In its detail superstructure could often be reassuringly domestic, its sheltered corners a verandah-like interface between the maritime world outside and the padded and panelled rooms within. In port the open decks of earlier liners allowed farewells to be extended through a thousand streamers as the ship pulled away from the quay. Stockholm still has well-maintained steam passenger vessels from the latter half of the last century operating alongside modern ones from its quays, and such steamships retain that intimate scale of design, along with extensive use of wood panelling and trim, which has been lost with the introduction of welded construction. Such vessels seem almost a Scandinavian chalet at sea, furnished as they are with wooden doors, decks and handrails, elegant wood-trimmed sash windows and curtains, and a general air of seagoing domesticity; in such a context comparisons between ship and land architecture seem all too easy, but these vessels are also sturdy craft designed for very specific waters and traffic. Even small prewar cargo vessels featured extensive use of wood in their bridge structures, and Dutch vessels (sometimes family-owned and -operated) are often exemplary both in the simple elegance of their design and their standard of maintenance.

The bridge area of the superstructure retained longest the use of wood-panelling, brass fittings, wooden decks and – even on small and modest merchant vessels, such as the coasters illustrated here – a general air of pleasure in handmade traditional materials, which owed much to the era of sail, and to its firm hierarchy of control. Other small cargo vessels trading in the 1960s had all-steel bridges which were cleanly innovative in design. Today's superstructures on cargo vessels are strictly functional, and any trace of exterior wood trim survives only on private yachts and some luxury liners. As Tony Lane wrote of traditionally-styled vessels in 1986:

When the last went, an easily paced and enjoyable style of living went with them. This was the end of the smart ship with stoned and bleached wooden

*FROM ABOVE: Small cargo vessel loaded with containers, Tilbury. Such ships can also carry a wide range of cargoes; The traditional colours of ship architecture – bridge and deck of the 1955 SS Shieldhall at Southampton; Flatspace architecture – a view of Tilbury container port*

*FROM ABOVE, L to R: Fine hull at the waterline, but all square on top – the 1985 container ship Nedlloyd Holland; The era of open deckspace – the 1949 P&O liner Himalaya; the passenger steamer Blidösund, over 100 years old and still working at Stockholm; Clean lines and a homely touch – a Dutch coaster at Aberdeen, 1963; The passenger steamer Norrskär and modern equivalent at Stockholm, 1995; Legacies of greater ships and grander days – the 1936 motor coaster Sincerity displays her wooden bridge front at Exmouth in 1963*

decks, varnished taffrails and fancy cordage on the accommodation ladder. Lingering in these and other liner-trade ships were many of the clipper-ship practices of decorative display.[18]

The bridge of a ship has design significance for a number of reasons: the visible and actual control centre of the whole vessel, its importance was traditionally signalled in the vessel's overall appearance by its need to be, in every sense, in a commanding position. Wood and brass trim aside, the priorities of good visibility and prompt access to modern navigation, control and safety equipment made the bridge responsive to changes in ship design and technology. A comparison of several twentieth-century bridge designs illustrates how both external modelling and internal layout reflect progress from what was essentially a modest and largely open visual command position derived from the traditional sailing vessel, to a fully enclosed one which bears comparison to an airport control tower. The bridges illustrated on page 106 are those of two Scandinavian coastal ferries and a Swedish icebreaker, and have in common the need to provide good all-round visibility even in the most severe winter conditions.

Gardner's bridge design for the QE2 combined practical considerations with awareness of the symbolic prominence of the bridge as the perceived centre of human command within the vessel's overall structure: he had it 'raised and projecting forward as brains and eyes of the ship', rather in the way that the sensing antennae of some insects both detect the way forward and are seen to be doing so.[19] Interestingly, P&O's Grand Princess will also have a distinctively protruding bridge module, which gives similar alertness to the profile. For all the developments in ship architecture, the design and positioning of the bridge in relation to the superstructure mass remains, together with funnel design, critical in ensuring that blend of purpose, alertness and balance which good design will bring. With the increasing adoption of an enclosed wheelhouse on the bridge after 1900 its roof eventually became the location of radio equipment with the superstructure even of small coastal and shortsea vessels acquiring a considerable and varied clutter which formed an important element in the overall appearance.

A good example of a vessel whose visible design was determined by a combination of operating conditions was the so called 'flatiron' collier which plied from Britain's North-East coalfields to power stations upriver on the Thames. Designed with minimal superstructure, such craft moved upriver (loaded, and so on maximum water draught) at high tide and could just pass under the Thames bridges; in ballast (and so with maximum air draught) they returned at low tide – and could again clear the London bridge arches. Early steam-powered flatirons had hinged funnels, as did many London river tugs; some later motor flatirons had telescopic funnels. These vessels have now disappeared along with London's upriver working power stations, although Bankside Power Station is to have a new lease of life as a modern art space for London's Tate Gallery.

Most dry cargo vessels up to the 1960s were distinguished by their array of derricks or cranes to handle varied items of cargo, with machinery often duplicated on quaysides of the period, but with bulk cargo or containers newer vessels keep decks as clear as possible to achieve fast turnaround times in port. Many new vessels possess a striking clarity of design with sound and confident proportions, but where earlier ships invited a lingering examination of small-scale deck

*FROM ABOVE: Break-bulk cargo – unloading crates, Aberdeen 1963; A study in simple patterns – the 1930 steam coaster* Barra Head *enters Lowestoft harbour, 1960; Designed for a very specific function – the South Eastern Gas Board flatiron collier* Croydon

*FROM ABOVE, L to R: Bridge front of the 1994 Hurtigruten vessel* Nordlys; *A clean and decisive all-steel bridge front, Aberdeen 1963; Superstructure and bridges of two modern icebreakers,* Otso *(nearest camera) and* Kontio, *Stockholm 1995; Flat plate economy – cargo vessel bridge, Bergen 1995; Contrasting bridge fronts of two ferries, Helsingborg 1995; The 1962 fast cargo ship* Glenogle, *which carried twelve passengers, shows her fine flare and sheer*

and superstructure detailing, the appeal of later vessels lies in larger, architectonic contrasts of mass. As with many modern buildings, any closer interest is likely to be in their sophisticated engineering and technology rather than in simply understood winches and derricks. Again, there are parallels with land architecture, both in scale and in the 'smooth', concealed nature of modern building technology compared with the visible fascination of, say, lifts in early hotels or swing-out cranes on nineteenth-century warehouses (now desirable 'period features' on luxury apartment conversions).

If hull form is largely determined by functional considerations, the arrangement and design of superstructure has always offered scope for individuality in styling. The cargo vessels operated by Alfred Holt of Liverpool, for example, were always unmistakable for their fine sheer, rounded bow front, pronounced bow flare, and towering funnels. Holts prided themselves on their high standards: their ships were not insured until the container era but their build standards were far in excess of all applicable specifications. Some shipyards discovered this to their cost, for the company exercised decisive influence in the design and building of all their vessels. By contrast, many yards today offer competitive prices only by severely restricting the extent to which the owner can effect variations on what is largely a fixed package; one British shipowner comments that with some Far Eastern builders the price may be competitive but the only alteration the shipowner can make to a vessel is the colour. Echoes of Henry Ford.[20]

Two new British shortsea tankers of innovative design for the British company FT Everard and Sons Ltd illustrate the internationality of ship design and operation: detailed aspects of hull design have been resolved by the specialist Norwegian firm Skipskonsulent; superstructure, gear design and crew accommodation are overseen by the company's own naval architect and Director at headquarters on the Thames, and the vessel is being built in Singapore.

Everards has for some years seen innovative design and technology as the best way to offset the inevitable rise in crewing costs, a policy in harmony with its determination to operate only with British crews trained in-house to meet the company's own high standards. Many shipping companies now operate with multinational crews for whom even communication in a common language can be a real problem, and this highlights the extent to which a ship is a floating world even more for its crew than for any passenger. One consequence that Everards has exploited in the positive sense is that, with proper consultation and planning, radical changes in work practices can be more readily introduced at sea than ashore; a new crew introduced to a new ship has every incentive to make their mobile working environment successful. The company's ships are modern and efficient, yet with purposeful style and appeal.[21]

Changes in crewing levels and in working conditions can have decisive effects on the appearance of a ship. The superstructure of modern passenger vessels has not only increased dramatically in height but is carried forward almost to the bow in order to accommodate all passengers above the waterline, and with a view of the sea as distant from engine noise or vibration as possible. Such development is possible through increased use of lightweight metal fabrication and has resulted in an overall smooth shape and profile quite different from earlier passenger vessels in which the visual punctuation of open decks, windows and

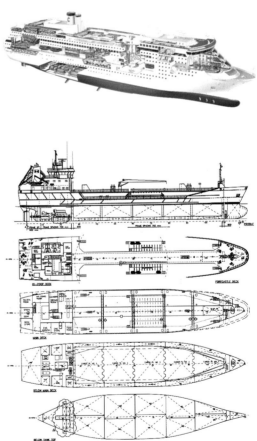

*FROM ABOVE: Older icebreaker at Helsinki, 1995; This cutaway impression of the Costa Victoria, now building for Costa Cruises, illustrates the complexity of such vessels; General layout of a new tanker currently building in Singapore for FT Everard and Sons Ltd, Greenhithe, designed by Skipskonsulent AS.*

porticoles was such a feature – exactly the earlier style the new Disney liners seek to evoke with their porthole-shaped windows and other 'period' features.

Less obvious influences can also determine ship profiles, as with the introduction of a complete deck level between the engine room and crew accommodation on cargo vessels so as to lessen engine noise (a consideration especially important with modern turbo-charged diesels). Again, some shipbuilding and shipowning companies ignore this consideration, and their crews have poor living and working conditions as a result. The problem of noise is also especially severe with modern icebreakers, whose role demands powerful machinery and also produces the endless din of metal hull against breaking ice; the very high superstructures of such ships is due largely to the need to provide bearable living conditions for their crews. On many ships this extra space is devoted to changing, recreational and store rooms, 'unnecessary' in the strict sense, but valuable in attracting quality crew members on vessels where numbers are planned to a minimum, and also in encouraging the crew to value and respect their working and living environment.

Exterior design often reflects interior, as well as exterior, requirements. On some vessels Everards moved exhaust uptakes to each side, partly to lessen noise, but this restricted space available for crew accommodation and also led to the risk of damage to funnel casings from high wharves when the vessel leaned inwards on a falling tide, or with cargo discharge; later vessels have a single funnel at the stern, while a single free-fall and enclosed lifeboat replaces side-mounted boats on davits for the same reason. Everard's 1990 tanker *Agility* also dispensed with the traditional bridge wings to achieve a slimmer superstructure inset from the ship's sides in order to minimise damage when loading and unloading. In the broader context of architecture, concerned as it is with people, it should also be noted that moving all superstructure to the stern had very immediate social effects for the crew:

> Whatever it did to aesthetics, the all-aft principle of construction . . . changed entirely the social structure of tankers, putting the engineers and navigators together, with the chief engineer and the ship's captain finally and unequivocally sharing status and the same deck.[22]

## IV Funnels

The funnel is the classic maritime motif of our day, not least because few vessels today really need one: prominent, distinctive, colourful and fun it is also essentially nostalgic. Yet despite technological advances, in most instances funnels still serve an important role in the design of a ship. Cruise liners have their own agenda, as we have seen, but in cargo vessels the funnel is the only design element that has survived an otherwise single-minded following of economic priorities by ship designers and owners. When questioned on this point as a naval architect, Marshall Meek sensibly observed that a funnel casing is a neat and tidy way of grouping the various pipes and vents on a ship, as well as creating useful storage space at deck level. He could not resist adding, however, that 'a ship wouldn't be a ship without a funnel'. The funnel is sometimes added as a complete unit late in fitting out, but its harmonious relation to other elements of superstructure design is of great importance; it must never look as though it has been added.

*FROM ABOVE: The 1990 shortsea tanker Agility, owned by FT Everard Shipping, Ltd, dispenses with traditional bridge wings to achieve an integrated wheelhouse/bridge structure which is also less liable to damage when the vessel heels during cargo operations; Icebreaker superstructure, Stockholm, 1995*

The two photographs of the 1962 liner *Northern Star* fitting out and on trial illustrate the process and the finished product.

One consequence of the motorship needing only a large exhaust pipe was that ship designers were able to indulge their fancies in shaping the funnel casing more or less to any shape they wished. A common tendency was to apply excessive streamlining to funnels on otherwise quite staid cargo vessels, with sometimes unfortunate results for the overall coherence of profile; if a cargo vessel has vertical masts for good practical reasons there is little sense in having a dramatically 'swept' funnel profile, unless the designer exploits this as a counterpoint: 'the disposition and shaping of the main masses [is] something which is dictated by service requirements'.[23]

The writings and drawings of Laurence Dunn over the years show clear thinking: he noted in 1966 that 'as cargo handling gear becomes heavier and more conspicuous' a rethinking of superstructure shapes was needed if 'overall balance of profile' was to be achieved.[24] As we have seen, some gear would disappear altogether, and he was also quick to see the need for new ways of thinking about gear in relation to the aesthetics of ship design:

> a range of new fittings has appeared which are architectural features in their own right, so that the shapes for hull, superstructure, etc, which were formerly desirable and appropriate, are now no longer so. To meet these new strong shapes sweet curves have to give way to greater severity. There can be no happy relationship between the old and the new.[25]

As Dunn notes in his paper, square funnels are nothing new, having been introduced as a distinctive design feature by Messageries Maritimes before the Second World War, but today such funnels are the norm because they are cheaper to construct and fit better on the deck plan of vessels which have themselves become more square in their sections, not least at the stern. The common use of a flat-section transom stern, which is cheaper to construct and offers more internal and deck space, encourages equally square superstructure sections.

A funnel-less design for merchant vessels was pioneered in 1912 by the first deepsea motor ship, the Dutch *Selandia*, which looked back to an earlier era in having sails on her foremast but whose mainmast carried exhaust pipes from her large diesel engines.[26] The absence of funnel here symbolised the modernity of the vessel's propulsion units, but the arrangement was also adopted by her later namesake in 1938. However, the pattern has not been repeated and it seems generally accepted that a funnel provides an essential focal point in a vessel's overall appearance, as well as a conveniently prominent location for the display of a company logo. The recent concentration of superstructure in a high stack at the stern of vessels has also seen a return of the twin funnel abreast format first used on Shell supertankers in the early 1960s. Considerable attention was paid to ensuring that the rear edge of the funnels echoed that of the superstructure – Laurence Dunn felt that the 'boldness of the bridgecurve' gave 'overall dignity' to this design, which was also helped by the deep white-painted hull section at bow and stern which both drew the design together and gave the vessel a deceptively low looking hull when laden.

It is rare today for funnels on a cargo vessel to play the decisive role in design as they did on *Serenia* and her sisters, or on the earlier P&O liners *Orcades*

*FROM ABOVE: The complete funnel of the* Northern Star *being lowered into position at the Vickers yard;* The Northern Star *on trials, 1962*

FROM ABOVE: Funnel casing on the ferry Pride of Bilbao;
Funnel and superstructure on the 1993 Hurtigruten
passenger vessel Kong Harald; Funnel and dome on the
Danish ro-ro passenger and vehicle vessel Crown of
Scandinavia

(1948), *Oronsay* (1951) and *Orsova* (1954), whose 'Welsh bonnets' sought to solve that perennial problem of how to disperse exhaust gases and smuts clear of passenger decks. More common is a token outline composed of latticework which clearly reveals the exhaust pipes within or one which features exhaust pipes protruding from the top. Even so, careful planning of design and deckspace enables the essential ducting functions of the funnel to become an architectural focus in its own right. An aerial view of the 1995 liner *Crystal Symphony* demonstrates that a funnel can serve several of the purposes discussed and still form an attractive top to a clean and tapering stern deck arrangement. For the present, at least, it seems likely that funnels of one shape or another will continue to serve a range of functions, and that no self-respecting ship will be seen without one.

### V  Rigging and Gear

On sailing ships rigging supported masts and controlled sails, the power source, as did much other related deck gear; sail power was complicated, labour intensive and dangerous. Rigging later supported the tall funnels of steamships as well as their masts, and some early steamers also carried a sail rig. Light rigging was needed for signal flags, while later came electronic gear and aerials for shipboard radio, radar and satellite navigation and communication. Other forms of working rigging were needed to control and operate the complicated cargo-handling gear that most ships carried, for originally all such cargo handling was done by manual labour using minimal onboard gear such as a pulley and rope tackle. Well into the 1960s many vessels were still handling cargo by winch and derrick, a decade after palletised and containerised cargoes pointed the way to present methods. In many respects cargo-handling gear itself changed remarkably little from the nineteenth century until the advent of container traffic: a series of derricks hinged at the base of the masts and equipped with pulleys driven by deck-mounted steam (later electric) winches functioned as primitive swivelling cranes for lowering hooks or nets into the ship's holds and depositing cargo onto the quayside, or vice versa.

In the age of break-bulk cargo, rows of quayside cranes would also work the ship but some vessels, such as those in the worldwide charter trade, were fully equipped to handle their own cargo as this might be necessary if the ship visited small ports, or those poorly provided with shore gear. Some vessels are still so equipped, including some container carriers, although with cranes rather than derricks, but it is increasingly common for ships to operate as cargoliners on fixed international routes or in bulk trades, and so not to carry such gear. With the increasing reliance on gearless vessels, both deepsea and shortsea ships have less of that independence of operation traditionally (although often wrongly) associated with the seafaring life, and are ever more dependent upon shore facilities. Turnaround times for container vessels are now counted in hours, whereas even in the 1960s a smaller ship (with a larger crew) might have spent over a month finding and loading cargo on some world routes. All ships now spend far more of their life at sea, but the price has been an acceptance that ships are but a unit in a larger operation rather than a free-moving entity. In reality this was always the case, but mechanised cargo handling and unitised cargoes have made this very apparent.

Some vessels always carried only minimal gear because of their specialised trade; tankers need only a small derrick to handle fuel pipes or to swing out a workboat. Passenger vessels, and especially those running between fixed ports, only ever needed gear to handle passengers' belongings, although in an earlier era this would have included some cars to be stowed in the hold. In general, passenger liners are characterised by the mass of their accommodation super-structure, punctuated by rows of portholes or windows, and by the larger win-dows of public rooms, but only marginally by the deck-level openings and der-ricks needed for cargo access. Modern cruise liners, whose passengers bring only hand luggage for their holiday trip, take this tidiness of profile to the limit.

At the other extreme, the seaborne export of heavy engineering units such as steam locomotives, or even finished small craft such as ferries or tugs, led in the period after the First World War to the development of merchant vessels in which cargohandling gear assumed major importance in both function and appear-ance. Much of this export trade was from Europe to the colonies, or former colo-nies, which had only limited port facilities and such vessels needed a lifting capacity of several hundred tons. The development of the Stulcken derrick greatly increased the lifting capacity of such ships, while the demands of the Second World War, and later of the postwar oil boom in Europe, resulted in ever more specialised vessels, from submersible oil platforms themselves to supply and rescue craft. All these vessels needed highly specialised gear in order to execute their tasks, many of which were quite new to the shipping world.

Such changes in the technology of shipping have profoundly affected not just the design of ships' gear but ships themselves. As hull design has shown, how-ever, some aspects of traditional ship design have not and cannot change be-yond a certain degree, although entirely new solutions such as hovercraft, hydro-foils, catamarans, and current new proposals for fast cargo vessels, are always possible. Yet an anchor today, of which there are many varieties, would no doubt be recognised as such by the sailors on vessels recorded in the Pharaohs' tombs; possibly an anchor winch, too. Apart from the largely unchanging design of sails, fishing nets and tackle, much post-sail gear such as ventilators, pulleys, cleats, bollards and even hatchcovers, is necessarily simple in design, construction and operation. Unless fully mechanical systems of docking and mooring are devised it is likely that such effective methods will remain in use, even on larger vessels.

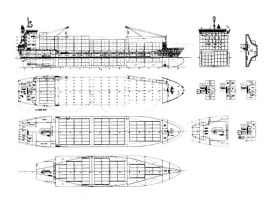

Many devices, and especially those employed in unchanging operations of seafaring, such as general cargo handling and docking, are also unaffected by scale and are essential to the largest and smallest vessels alike. Both a rowing boat and a supertanker need to be moored securely, and for efficiency a hemp rope or steel hawser is best handled by few people; on the supertanker an initial line attached to the rope may be fired from a powergun and a powerful winch will then haul in line and rope and take the strain, but an understanding of momentum, gravity, friction and tension – as well as a good eye and quick reflexes – are always essential. Unique to ship architecture is this recognisable similarity of such basic technological elements as bow, rudder, anchors, bridge, funnel and stern to designs of a hundred years before. No land-based architecture of similar scale and subject to such major technological change could retain such tradi-tional elements without being consciously reactionary or inefficient – or both. Yet

*FROM ABOVE: Carefully shaped and integrated superstruc-ture design on the 1995 cruise liner* Crystal Symphony, *owned by Japan's NYK Line and built by Kvaerner Masa-Yards; This Seacat trimaran brings age-old Pacific under-standing of hulls and outriggers to modern ferry technology; Plan for a container carrier equipped with cranes by Skipskonsulent AS*

FROM ABOVE: This view of the Swedish icebreakers Otso and Kontio shows the funnel element well integrated into superstructure, Stockholm, 1995; Traditional lines of the prewar steam drifter Kestrel; Painting the Holland America Maasdam, built in 1993 by Fincantieri, Stockholm, 1995

as maritime disasters show, when even the largest vessels are in danger of wreck the skills needed to save them are all too familiar; nor are such operations helped by the large scale of some modern vessels, and by their frequent dependence, for reasons of cost, upon single engines, rudders and propellers:

> Such an outlandish hull manoeuvred by one propeller and a single rudder is on the face of it ludicrous: relatively low drag allows supertankers to roll easily through the water, so the single-screw system is adequate for thrust, but it is wholly inadequate for emergencies, and hopeless as a means of breaking the considerable momentum of the ship in a hurry.[27]

The importance of human scale, both in the operation of gear and in measuring its diversity of shape, size, colour and function against the larger, unbroken masses of a vessel's hull and superstructure is crucial to a ship's overall design. On small coasters, fishing vessels and tugs scale is still sufficiently human for the whole vessel to be taken in at a glance, and for orders to be shouted from the bridge accordingly. Undeniably something is lost when, as on many large vessels today, crew members communicate by radio telephone and closed circuit television cameras, providing the bridge with information on different areas of the vessel, or on a quayside hidden from view by bulky superstructure or bow flare.

Technology increasingly renders ships' gear more efficient and sometimes, as with on-board gantries and container cranes, makes gear itself a major element in the vessel's profile. Conversely, some gear is invisible: on fishing vessels the necessary winches, booms and nets have become more prominent with more mechanical fishing methods, the vessel itself sometimes being significantly adapted in the process. Yet even a small vessel will now have a sonar device mounted underwater for scanning the sea bed and locating fish shoals. Vessels from coasters to cruise liners are now fitted with transverse bow thrusters for manoeuvring in port, but such important design developments are visible only when the ship is in dry dock; the same is true of the powerful and highly sophisticated propulsion and steering methods fitted to many tugs, vessels whose active element is very much beneath, not above, the waves.

If the role of gear on merchant vessels varies, rigging itself is now minimal in the strict sense. Hoisted signals of various kinds still play an important part in navigation, and all vessels need rigging sufficient to display these; radio aerials are essential but are now supplemented by radio communication and satellite navigation equipment. On a large modern vessel the area atop the wheelhouse carries a formidable array of aerials, scanners, globes, domes and antennae, but steel masts (where still fitted) and the various deck-mounted container transporters and gantries do not need the supporting rigging essential to earlier, more lofty and fragile masts and spars.

### VI Cargoes

The cargo or cargoes to be carried by a ship, together with the routes to be plied, are the most important factors in determining its size and shape; this is as true for the passenger liner as for the tanker. This section's title is stolen in one sense from John Masefield's 1910 poem 'Cargoes' (famous above all for its 'dirty British coaster'), but the poet is more interested in the associations of each cargo than the ship designer and shipowner who must think of costs.

Few ships today are built for the tramp trade, in which a ship essentially plied for hire and where cargoes, even perhaps for a particular voyage, were sought and fixed at short notice by a shipping agent. In a competitive world ships still find cargoes in this way, but increasingly the investment a shipowner must undertake in ordering a new vessel (which may not be delivered for two or three years) makes it prudent to have a clear understanding of that vessel's likely trades and profits, at least in terms of recouping the outlay and any associated borrowing. Ideally a cargo ship will be built for charter to a company that needs just such a vessel to ply a particular trade, and specifications will then be agreed in close consultation with that company. But all ships, whether passenger or cargo, need to retain maximum flexibility in their design in order to cope with unforeseen changes in market conditions. Such flexibility may affect the shape and equipment of the vessel very considerably, and a wise shipowner will know what must be incorporated in the ship to allow for the future possibility that it may trade on a different route, and with different cargoes.

Some cargoes are so specific in their requirements that any such flexibility is possible only in routes to be operated, not cargoes carried. A good example is liquified gas, which is carried in cooled form within pressurised spherical containers located within the hull proper of the ship. The construction quality of these containers must be rigorously controlled, and in their dramatically simple form they illustrate very clearly the primary role of the ship as container and transporter of materials. The construction plans of similar vessels show alternative methods of locating these inherently strong, rounded shapes within the square-section dimensions of a ship's hull.

In the days of break-bulk cargoes the planning, supervision and stowage of many hundreds of packages containing materials of different weights and sizes (and smells) was a highly skilled task, as was the labour intensive business of stowage itself. Delicate cargo had to be protected, care taken that foodstuffs were not contaminated by odours from leather or chemicals, and everything stowed away on the various decks of the hold in such an order that both the stability of the ship and the order of eventual unloading *en route* (which, along with fuel consumption, had its own effect on stability) were taken into account. Stability remains a crucial consideration for all ships, and is not simply a matter of locating as much weight as possible low down in the hold; such loading can cause a ship to swing like a pendulum, and allowance must also be made for a large vessel's tendency to hog and sag in a long sea. Containers of unit size, along with onboard specialised software, have greatly simplified and speeded cargo loading calculations, but some basic problems will always remain. Among the most problematic cargoes are those which are inherently dangerous, such as petroleum and gas, or those which are liable to shift during transit, especially in heavy weather; ore, coal, stone and grain (which also swells if water reaches it) are among these. Such bulk cargoes are quick and easy to load and unload with modern machinery, but some ores are so heavy that only a small amount constitutes a full cargo by weight. Care must be taken in the design of ore carriers to ensure that the bottom of the hold is sufficiently strong and shaped to prevent cargo from shifting; some have hopper-shaped holds with conveyor belts running along the bottom of the vessel, while others carry unloading gantries and conveyors on deck.

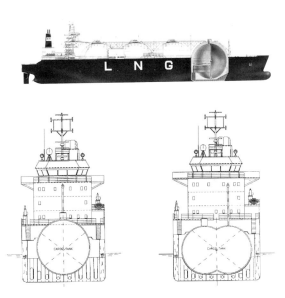

*FROM ABOVE: Monumental scale – Aquitania funnels and officer 1914; Profile plan for a liquified gas (LNG) carrier by Kvaerner Masa-Yards, Finland; Transverse section of a liquified gas carrier (showing two possible configurations of the tanks within the hull) by Skipskonsulent AS*

113

*FROM ABOVE: Wartime Landing Craft in action. Their very basic, but nevertheless revolutionary, bow ramps inspired ro-ro ferries in the postwar era; As architecture most ships recycle themselves within thirty years – the cargo vessel* Ziyak *loads scrap at Tilbury; Handling break-bulk cargo on pre-container vessels entailed long and hard work stowing small packages on different decks. The vessel is the 1944* Talthbius

With bulk cargo the movement involved in handling at the point of shipment has always been basically vertical: grain or scrap is emptied into the hold, to be sucked or lifted out for discharge; oil is pumped in or out. With break-bulk cargo the relatively small hatches on vessels involved much manhandling of packages from the central loading shaft, into which derricks could lower cargo, to the furthermost corners of the hold. The major change brought by full-scale containerisation was that all shipboard cargo handling was simplified to the vertical plane, with containers slotted neatly and securely into guides which form part of the vessel's very structure. In port the custom-made container gantries reach the full width of the vessel to carry out loading and unloading, operating as a direct link between the ship itself and road or rail links on the quayside. Plans for such vessels demonstrate the manner in which every aspect of their design accommodates itself to the economic requirement for cargo to be stowed and moved quickly and efficiently, with superstructure above the deck and engines below it all located so to minimise interference with the container cargo.

The compositional rhythms of such vessels are far removed from those which carried all their cargo within the hull: all is now dominated by the containers themselves and their often multicoloured patterns and changing shapes, but this visible cargo impresses by its very mass and also contrasts with a hull which needs to remain as slim and speed-conscious as possible. The design of the hull is for fine horizontal water penetration, that of the container stowage for vertical mechanical handling, and a container ship demands these two very different 'readings' in order to be understood as a functional construct. For some this is an impossible task and what they term 'box ships' are dismissed as unworthy of the name 'ship' at all. In buildings, too, acceptance of a new verticality is required to understand how urban constructions function, and container vessels in some respects parallel this development rather than, as in the past, contrast with it. A modern container vessel is an impressive sight by any standards, its form dictated just as unrelentingly by the demands of its cargo as was that of the *Cutty Sark*; pressed bales of wool and chests of tea were, after all, precursors of today's standard containers.

Some ships are virtually devoid of gear not because (as with bulk shipments or containers) the handling is done by shore-based equipment but because their cargo is self-propelling. Roll-on roll-off ferries are among the most important developments in modern shipping, bringing maritime and road transport into close integration. These vessels derive from wartime initiatives in bringing equipment and stores by sea right onto the beaches, from where they could be taken to bridgeheads inland. The North African and Normandy D-Day landings required the design and building of special craft which could carry troops, motor vehicles and supplies right onto the beaches. The Landing Craft devised for these uses was the forerunner of postwar roll-on roll-off commercial ferries, and a related supply vessel was the 'Chant', some forty of which were tankers and some twenty dry cargo. These motor vessels were of simple flat-plane construction and were designed to be welded up in units at factories inland and then finally assembled at various British shipyards. The vessel's flat bottom and angled stern were designed for easy beaching, and many of these craft were still in service around the coasts of Britain well into the 1960s; abroad, some still are.

Some cargoes carried by ships are so large and unwieldy that they will go neither in the ship nor on deck. To cope with such loads, which can often include other ships as well as oil and gas drilling equipment, vessels have been developed which can partly submerge themselves and then rise by buoyancy under the load which has been floated into position over them. Other loads can be craned directly onto their extensive flat surface and then secured for transit.

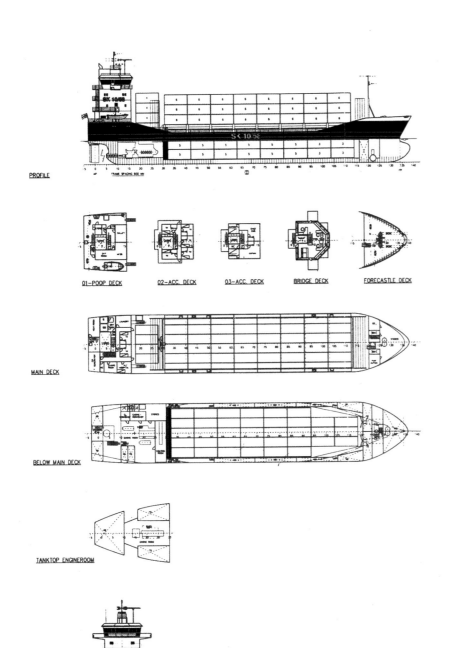

*ABOVE: The 1983 semi-summersible Wijsmuller heavy lift vessel Mighty Servant. Such specialised craft flood tanks to float beneath their load, then pump tanks to lift it by sheer buoyancy; BELOW: A wartime Chant tanker, showing the deliberately simplified flat-plate construction. Many were still in use in the 1960s; LEFT: General arrangement of container vessel by Skipskonsulent*

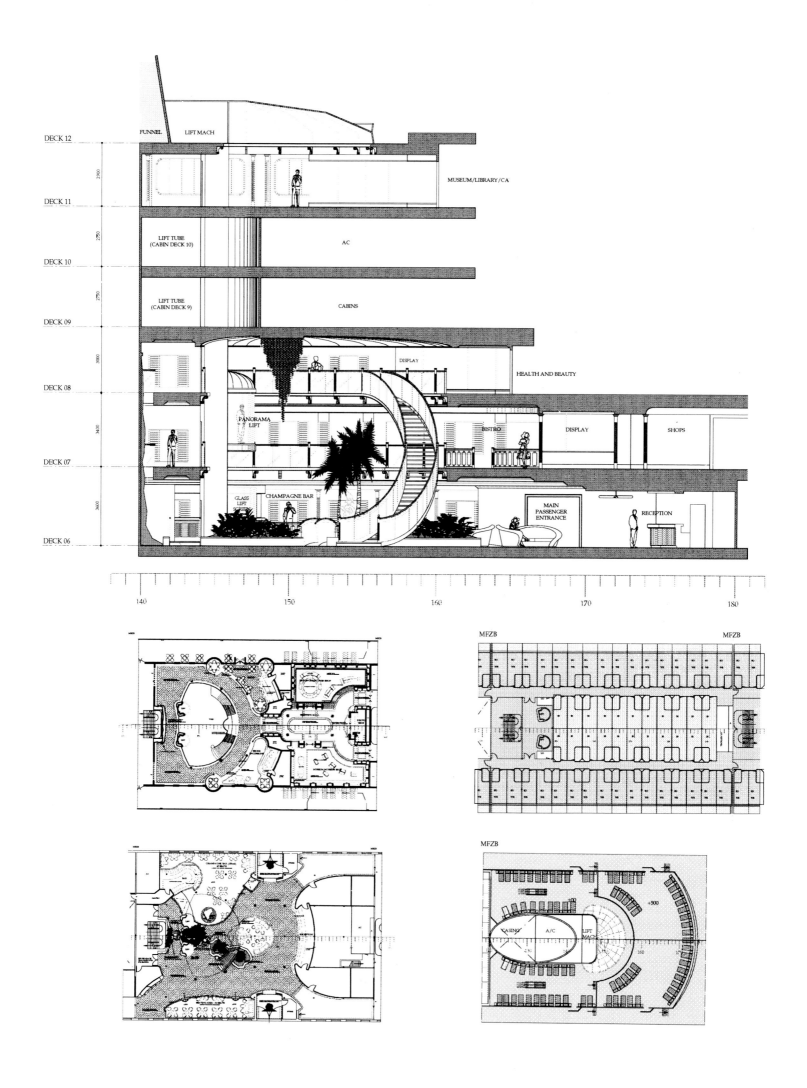

DECK 12

DECK 11

DECK 10

DECK 09

DECK 08

DECK 07

DECK 06

FUNNEL    LIFT MACH

MUSEUM/LIBRARY/CA

LIFT TUBE
(CABIN DECK 10)

AC

LIFT TUBE
(CABIN DECK 9)

CABINS

DISPLAY

HEALTH AND BEAUTY

PANORAMA
LIFT

BISTRO

DISPLAY

SHOPS

GLASS
LIFT

CHAMPAGNE BAR

MAIN
PASSENGER
ENTRANCE

RECEPTION

140        150        160        170        180

MFZB                MFZB

MFZB

CASINO    A/C    LIFT MACH    +500

# CONCLUSIONS

*To like ships is first and foremost to like a house, a superlative one since it is unremittingly closed, and not at all vague sailings into the unknown: a ship is a habitat before a means of transport.*

*Roland Barthes,* Mythologies, *1957*

*The ship, powerful and mysterious construction, a being possessed of a life of its own, floating building and city; a dreamlike image such as the Rex in [Fellini's]* Amarcord, *unreal apparition, sparkling with lights, which takes shape and then vanishes into the night: a ship is all this.*

*Donato Riccesi, 1985*

In *The Black Atlantic*, his study of 400 years of transatlantic cultural relations, Paul Gilroy suggests that the two most important cultural carriers in history have been the slave ship and the long-playing record.[1] The modern cruise ship, built perhaps in Finland, owned in Britain, registered in Nassau and operating out of Miami with passengers of every nationality, suggests ways of extending Gilroy's suggestion; the problematic assumptions informing Western attitudes to supposedly exotic (and usually poorer) destinations have characterised international cultural studies certainly since the publication of Edward Said's *Orientalism* in 1979; the modern cruise market and its vessels offer a rich area for study.

One visible influence of hotels and shopping malls on cruise liners has been the atrium, an architectural device indirectly borrowed from the nineteenth-century glass-roofed shopping gallery by hotels of the 1970s onwards.[2] The Dundee-built *Zealandia* of 1899 featured a glass dome some eight feet wide over its saloon, or possibly stairwell, and later liners had employed steel trellis girders to support large skylight structures.[3] In one sense an architectural descendant of glass-roofed courtyards and shopping galleries, in a shipping context the atrium is the creation of Njärl Eide, although some liners in the last decade of the nineteenth century featured public rooms which were two decks in depth; Stephen Payne cites Cunard's *Campania* and *Lucania*, both of 1893, as early examples.[4] Today's sophisticated atrium, with computer-controlled lights, waterfalls and glass lifts echoes the way Las Vegas hotels and casinos employ high technology and a themed approach to accommodation and entertainment. The projected Tillberg liner of 'classical design' concealed within its carefully-shaped profile an atrium 'three decks high' with 'two glass elevators, a waterfall and a tropical landscape with a bridge over a stream coming from the waterfall'.[5]

A major breakthrough was the 1987 cruise liner *Sovereign of the Seas*, built for Royal Caribbean Cruise Lines, which included an atrium by Eide spanning five decks, and this company has made both the atrium and the related 'Viking Crown'

*OPPOSITE: Proposed atrium for a post-modern 'classical' liner by Tillberg Design*

lounge – which is incorporated as part of the forward funnel structure and serves as roof to the atrium – part of its company identity; when *Nordic Prince* was sold the company required the Viking Crown lounge to be removed by the vessel's purchaser.[6]

Many vessels have increasingly exploited the growing height of their super-structure to create glassed-in verandahs offering spectacular views from public rooms; the sea itself becomes for much of the voyage merely a view from a luxurious floating hotel, or resort city.[7] The futuristic plans by Norwegian ship-owner Knut Kloster, who successfully transformed the transatlantic liner *France* into the cruising liner *Norway*, include the 250,000-ton *Phoenix World City*, three times longer than the largest cruise liner and (in the best tradition) 'almost twice the height of Nelson's column'. The vessel would have to be launched in three parts and joined together later: 'Its specifications will include pools, conference halls, a shopping mall, a television station, casino, art and music studios and sports centres. There are plans for a "university" catering for students . . .'[8]

At present, however, cruise vessels can best achieve this sense of space through the atrium, which serves as a focal point for some of the most striking aspects of cruise ship design generally. The central importance of the concept has been accepted by cruise ship operators and their passengers, and in arguing the importance of a similar space in cruise ship interior design Eide draws analogies with town planning and the role of 'the central plaza, a rendezvous where people can congregate and socialise'.[9]

Viewed within the tradition of passenger ship development, the one spectacu-lar effect of the atrium is that it establishes a major public space linking several of the horizontally-defined decks which traditionally formed a distinctive feature of these vessels, externally and internally. Decks once reflected an equally clear social stratification determined by cabin price. On cruise vessels such custom is truly captive to an extent that might echo the nightmare urban landscapes of modern Los Angeles explored in Mike Davis's 1990 study *City of Quartz: Excavat-ing the Future in Los Angeles*.[10] Jencks has seen the convention centre as 'the building type of the nineties', just as the museum had been of the eighties; the modern cruise liner combines elements of both and may yet prove to be the perfect metaphor of post-modern culture.[11] The terrestrial world that passengers leave behind is increasingly violent and disturbing, and such social pressures reinforce the basic philosophy of cruising – that, literally and symbolically, it is more pleasurable to travel rather than to arrive.

The very operation of a SWATH [Small Waterplane Area Twin Hull] defies ex-planation by the lay person, and its designers accept the need to sell the revolu-tionary appearance of the craft rather than its novel operational principles. Once again, the visible yet passive element of ship design serves to express technical developments realised by its active, but invisible, aspects:

> The platform and the pontoons are very fascinating for shipbuilders, but they are not going to be that fascinating for the passengers. They do not understand such things as interaction between moving cylindrical bodies below water, and I do not know if they know about frictional area and things like that either . . . But they are going to be attracted by the new look of the SWATH ships and we are going to play on that.[12]

He added that in designing modern vessels it must be remembered that, while more traditional passengers still 'dream of the *Queen Mary*' and 'do not want to have the new P&O ship' [the 1990 *Crown Princess*] other passengers will relish being able to 'tell the neighbours they have been on a twin-hull ship instead of the normal monohull ship'.[13] Such issues may indeed be 'far away from Naval Architecture' as traditionally understood, but are nevertheless central to the function of ship architecture as explored here.

The unplanned glimpses offered by adjacent themed lots in Las Vegas (South Sea palm trees visible through the columns of Caesars Palace) have been relished by Alan Hess in his appreciation of the city's architecture; cruise vessels afford similar opportunities for disparate theming, the one proviso seemingly being that there should be no maritime aspect, something that Las Vegas can and does exploit in its desert setting.[14] Such visual conceits take post-modernism-as-pastiche to new levels but such rich randomness, combined with vessel exteriors that increasingly owe as much to contemporary design in other transport and construction fields as to any 'nautical tradition', contradict Jencks's comparison of Foster's Sainsbury Centre with an 'architecture of ships and holidays, of activity liberated from the weight of a past felt to be crushing'.[15] On the contrary, such nautical imagery as there is in Foster's building owes more to the rigging of conventional interwar ships, traditional Cunarders and Victorian cast iron and lattice piers – the 'weight of a past' with a vengeance – than to modern shipping technologies.

That our post-modern society ignores the merchant fleets upon which its global transport systems and wealth depend is a cultural phenomenon that has been little explored, by architects above all; speculation stops at the shore. There is both scope and need for those with commercial and design interests in modern shipping to stimulate informed debate in areas where ships have every claim to be in the forefront of architectural practice. For many land architects it will be a challenge to engage with ship architecture on a scale and with a detail which precludes lazy general references to 'nautical' details which are themselves all too often cliches. The scale and the transience alike of modern ships poses a challenge to maritime historians and to maritime museums: whole generations of modern vessels disappear with scant attention paid to them outside the pages of naval architectural or enthusiast journals, while many who might redress this situation still display nostalgia for sail and steam.

This book is an outsider's view of shipping, and of the forms modern trade has produced. At one level it seeks to reclaim ships as architecture; self-propelled buildings on the sea. It may be that in the future the imaginative significance of ships, shaped as they and their ports increasingly are by forces largely divorced from public gaze and involvement, will be very different from in the past; if so this will entail significant shifts in the assumptions of architects and maritime museums alike. One thing is certain: until the form and meaning of modern ships occupies a more central place in architectural and cultural debate this will remain significantly impoverished. The sea is both a crucial, and crucially fragile, dimension of our world; our cultural apprehension of that element and its trades must complement financial and commercial ones if we are to build on the sea with informed confidence and understanding.

**Ship,**
**house on the water,**
**I salute you**

Edward Kamau Brathwaite
*Islands*, 1969

# NOTES

**CHAPTER 1**

1 I am indebted to Gianfranco Bertaglia, William Everard, Marshall Meek CBE and Fred Walker for invaluable insights into the professional theory and practice of naval architects. Subsequent unattributed comments in this text by Meek are from my interview with him at the Royal Institution of Naval Architects, London, 13 September 1995.

2 The maritime life and culture of the Scottish East Coast is traced and illustrated in Peter Anson's *Fishing Boats and Fisher Folk on the East Coast of Scotland*, JM Dent (London) 1930; reprinted 1971.

3 For useful commentary on the changing history of ports see the studies by William J Fishman et al, Gordon Jackson, Chris Ellmers and Alex Werner, and also Dockland (various authors), in the Bibliography.

4 Some of the better books in this field are listed in the Bibliography.

5 Sir Norman Foster, 'Boeing 747' (designed by Joseph F Sutter) in *Building Sights*, Academy Editions (London) 1995, p53.

6 Foster, p55.

7 Lord Byron, *Childe Harold's Pilgrimage*, Canto I (1812), Stanza 179.

8 In Alan Hess, *Viva Las Vegas: After-Hours Architecture*, Chronicle Books (San Francisco) 1993, p7.

9 Sir Hugh Casson, 'A Ship is An Island', Special Issue of the *Architectural Review* (to celebrate the QE2), CXLV:868, 1969, pp399-419; reference on p405.

10 Casson, p407.

11 Casson, p405.

12 Donato Riccesi, *Gustavo Pulitzer Finali, Il Disegno della Nave: allestimenti interni 1925-1967*, Marsilio Editori (Venice) 1985, p58. Riccesi's book is an incisive and well-illustrated work on ship architecture, unfortunately not available in English. Translations here are my own, with any problematic phrases also given in the original Italian.

13 Paul Fussell, *Abroad: British Literary Traveling Between the Wars*, Oxford University Press (New York) 1980, p45.

14 Ted Scull, 'Disney News' in 'Cruise Ship Review', *Ships Monthly* (Burton-on-Trent) Vol 31, No 2, February 1996, p29. In a 1995 e-mail Alan Hess observed of Disney's involvement with ships that for many years they ran the *Queen Mary* at Long Beach; that venture proved unsuccessful but the failure in running a real historic ship is being put to good commercial use in planning a 'retro' one.

15 Joseph W Konvitz, 'The Crises of Atlantic Port Cities, 1880 to 1920', *Comparative Studies in Society and History*, 33:2 (April 1994) pp293-318.

16 Reyner Banham, 'Flatscape with Containers' in Charles Jencks and George Baird (eds) *Meaning in Architecture*, Barrie & Rockliff: The Cresset Press (London) 1969, pp103-8. Reference on p107.

17 Jan Morris, 'Watching the Ships Go By', *A Tangled Life* (Arrow Books, London) 1990, p47, first published Barrie and Jenkins Ltd, 1989.

18 Allan Sekula, *Fish Story*, Witte de With, Center for Contemporary Art (Rotterdam) and Richter Verlag (Düsseldorf) 1995, p54.

19 M Christine Boyer, 'Cities for Sale: Merchandising History at South Street Seaport' in Michael Sorkin (ed) *Variations On a Theme Park: The New American City and the End of Public Space*, Hill and Wang (New York) 1992, p190. See also Kowinski, William Severini, *The Malling of America: An Inside Look at the Great Consumer Paradise*, William Morrow and Company Inc (New York) 1985.

20 See Gerald Silk, *Automobile and Culture*, Harry N Abrams Inc (New York) and The Museum of Contemporary Art (Los Angeles) 1984; Nick Georgano, *Art of the American Automobile: The Greatest Stylists and their Work*, Prion/Multimedia Books (London) 1995; also (more generally, and including ship design) Richard Guy Wilson, Dianne H Pilgrim and Dickran Tashjan, *The Machine Age in America 1918-1941*, The Brooklyn Museum in association with Harry Abrams Inc (New York) 1986.

21 See Francis D Klingender, *Art and the Industrial Revolution*, Paladin (St Albans) 1972 (first published 1947); also Ken Baynes and Francis Pugh, *The Art of the Engineer*, The Overlook Press (New York) 1981.

22 William Ivins Jnr, *Prints and Visual Communication*, Harvard University Press (New York) 1953 and Estelle Jussim, *Visual Communication and the Graphic Arts*, Bowker (New York and London) 1974.

23 Cervin Robinson and Joel Herschman, *Architecture Transformed: A History of the Photography of Buildings from 1839 to the Present*, The Architectural League of New York (New York)/The MIT Press (Cambridge, Massachusetts) 1987, px.

24 ibid, p58. This fine study contains a factual error which demonstrates the need to understand ship architecture in detail and in context: Herschman interprets sailing barges and lighters in an 1847 calotype by Fox Talbot of Hungerford Suspension Bridge on the Thames as 'the picturesque but inefficient traditional means of crossing the river' (p32). But see the valuable comments on the 1938 photograph of Rotterdam apartments and harbour by Jan Kamman (pp115, 116) and on the 1958 photograph by Walker Evans, 'Boat-and-davit on the *Liberté* portside' (pp140, 142).

25 Konvitz, p296: 'The industrialization of shipping, the expansion of ports, and the presence of considerable industry in waterfront districts had the effect of blurring those features of ships and ports which took the imperatives of seafaring into account (pp296-7).

26 Roger Quarm and Scott Wilcox, *Masters of the Sea: British Marine Watercolours*, Phaidon Press Ltd (Oxford) in association with The National Maritime Museum (Greenwich) and The Yale Center for British Art (New Haven, Connecticut) 1987; reference on p68; the comment is by Wilcox.

27 For an absorbing and wide-ranging study of our uses and abuses of the past in interpreting the present see David Lowenthal, *The Past is a Foreign Country*, Cambridge University Press (Cambridge) 1985.

28 For the seagull tendency see the article by Demetri Porphyrios on the contemporary British architect Rob Krier in *A Decade of Architectural Design*, Academy Editions (London) 1991: 'Aerial perspectives are a favourite device of this architect, because they are more effective in presenting the city as an entity, rather than the sum of individual parts'. (p27)

29 Robert Venturi, Denise Scott Brown, Steven Izenour, *Learning from Las Vegas: The Forgotten Symbolism of Architectural Form*, The MIT Press (Cambridge, Massachusetts) 1994 (first published 1972), p152.

30 Casson (1969) p400.

31 Quoted in Venturi et al, p136.

32 For a detailed illustrated account of the development and design of this element of ship architecture see GR Wilkinson, 'Wheelhouse and Bridge Design – A Shipbuilder's Appraisal' in the *Transactions of the Royal Institution of Naval Architects* (London), Vol 113, No 7, Jan 1971, pp237-81.

33 See Håkan Stenberg, 'Silja Line's popular promenades', *Maritime Transport International*, 1993, pp167-70.

34 This information came from Robert Tillberg. For a range of articles on the new *Oriana* see *Designs 95*, Plus 2 Ferryconsultation (Halmstad) 1995.

35 Peter Anson, *Harbour Head, Maritime Memories*, The Catholic Book Club (London) 1944, p145. A pen drawing of the *Duchessa d'Aosta* at sea appears (unidentified) in Anson's *How to Draw Ships*, Studio Vista (London) 1941, p50.

36 Venturi et al (1972) made the much-cited distinction between 'the duck' and 'the decorated shed' and commented: 'It is now time to re-evaluate the once-horrifying statement of John Ruskin that architecture is the decoration of construction, but we should append the warning of Pugin: It is all right to decorate construction but never construct decoration'. (p163)

37 Nöel Mostert, *Supership*, Macmillan (London) 1974, Introduction, pix.

38 Alain Corbin, *The Lure of the Sea: The Discovery of the Seaside in the Western World 1750-1840*, Penguin Books (Harmondsworth) 1995 (originally published Paris, 1988, as *Le Territoire Vide*), p16.

39 Sekula, p12.

40 I am indebted to Fred Walker for a usefully sceptical assessment of Brunel's engineering importance (as his record on costing projects was so bad).

41 John McKean, *Crystal Palace: Joseph Paxton and Charles Fox*, Phaidon Press Ltd (London), p41.

42 In *New Directions in British Architecture: Norman Foster, Richard Rogers, James Stirling*, Thames and Hudson (London) 1986, Deyan Sudjic comments: 'For Foster, Joseph Paxton's Crystal Palace was the greatest piece of architecture of the nineteenth century, and has still hardly been equalled'. (p39)

43 Loudon, as quoted in McKean (1994) p44.

44 McKean, p44.

45 McKean, p41.

46 'Transforming a Scandinavian Liner into an Italian Beauty', *Cruise and Ferry Interiors*, Contract Communications Ltd (London) 1996, pp87-9.

47 Comments and quotations here by Gianfranco Bertaglia, naval architect at Fincantieri and the original designer of *Crown Princess*, are from my interview with him in Triest, 31 August 1995.

48 In his Introduction to *Shipwreck* (with photographs by the Gibsons of Scilly) Jonathan Cape (London) 1974, the writer John Fowles comments: 'A wrecking sea is part of what we all dream ourselves to be every night . . . So the shipwreck is not only what we are thankful will never happen to us; it is also what we secretly want to happen, and finally to ourselves'.

49 See Marshall Meek, 'The First OCL Container Ships', *Transactions of the Royal Institution of Naval Architects* (London), Vol 112, No 1, 1970, pp1-41: 'The salient features of the hull form are the very pronounced flare of the bows designed to cast the seas clear of the deck of containers' (p5).

50 Meek, p1.

51 Sekula, p12.

52 Robert Kronenberg, *Houses in Motion: The Gen-*

*esis, History and Development of the Portable Building*, Academy Editions (London) 1995, p10.

53 Kronenberg, p91.

54 Jonathan Raban, Introduction to *The Oxford Book of the Sea*, Oxford University Press (Oxford) 1992, p18.

55 Sekula, p183.

56 Sekula, p50.

57 ibid.

58 Mostert, p71.

59 Friedrich Engels, *The Condition of the Working Class in England* [1845], translated and edited WO Henderson and WH Chaloner, Stanford University Press (Stanford, California) 1968, p30. As quoted in Sekula, p42.

60 Sekula, p51.

61 Gianfranco Bertaglia, 1990 press statement on ship design and culture entitled 'Progetto Nave' [Ship Design and Planning], 3 April 1992. My translation.

62 Quoted in Reyner Banham, *Design by Choice*, Academy Editions (London) 1981, p47.

63 I am grateful to Fred Walker for improving my grasp of these important technical issues.

64 Marshall Meek pointed this out to me. His paper (Meek, 1970) states: 'A feature that the designers found intriguing in these first container ships was that for the first time in dry cargo ship design the nature of the cargo is known precisely . . . Because containers are supported at their corners only, the positions of the ship's transverse strength members and transverse frame spacing are related directly to container length (p4).

65 Jonathan Raban, *Hunting Mister Heartbreak*, Collins Harvill (London) 1990, p29.

66 See Caroline Tisdall and Angelo Bozzolla, *Futurism*, Thames and Hudson (London) 1977, p7.

67 Le Corbusier, *Vers un Architecture* (1923) trans as *Towards a New Architecture* by F Etchells, John Rodker (London) 1927.

68 Sekula, p183.

69 Joseph Conrad, *Heart of Darkness*, Bantam Books (New York) p6. Conrad's story was first published in 1899.

70 Tony Lane, *grey dawn breaking: British merchant seafarers in the late twentieth century*, Manchester University Press (Manchester) 1986, p80. Lane's is a rewarding study by a former seafarer turned academic.

71 Lane, op cit, p81. See John Heskett, *Industrial Design*, Thames and Hudson (London) 1980, especially pp38-40. Heskett observes shrewdly of liner design: 'Their design related not only to essential considerations of seaworthiness and performance, but to the transport of passengers in surroundings providing comfort and reassurance . . . the relationship between form as mechanism or structure, and as decoration and symbol, is complex . . . (p40). See also Heskett's writing on liner interiors in his *Design in Germany, 1870-1918*, Trefoil (London) 1986.

72 Heskett, p10.

73 Raban, p7.

74 Sekula, p50.

75 Mostert, p238.

76 Hess, p33.

77 Margaret Crawford, 'The World in a Shopping Mall' in Sorkin (1992) pp3-30 and 233-6 (references on pp3 and 21).

78 Ship imagery persists in both the conscious planning of Las Vegas (the ultimate fantasy amid the desert sands) and in commentary on its architecture. Hess quotes Reyner Banham as writing in the 1970 *Los Angeles Times*: 'The Strip . . . is grandly a string of island palaces in a sea of dark, connected by a canal of leisurely automotive trans-

port' (p87).

79 Charles Jencks, *The New Moderns, From Late to Neo-Modernism*, Academy Editions (London) 1990.

80 Maurice Culot and Jean-Francoise Lejeune (eds) *Miami: Architecture of the Tropics*, Princeton Architectural Press (New York) 1993, p18. Intriguingly, this lovely book omits any discussion of the many cruise vessels based at Miami, although it does include a picture of one – and is alert to the role of the ocean in Miami's architecture and culture.

81 Joseph Farcus, Public Architect, Miami, extract from a 'Company Profile' faxed to me, February 1996.

## CHAPTER 2

1 Margaret A Rose, *The Post-Modern and the Post-Industrial: A Critical Analysis*, Cambridge University Press (Cambridge) 1991, pxi.

2 Joseph Farcus, 'Company Profile', 1996. Further unlocated quotations here are from this document.

3 Sekula, p106.

4 Sekula, p43.

5 Joseph Conrad, *The Nigger of the 'Narcissus'*, (first published 1892) as quoted in Raban, p42.

6 In the UK quality magazines such as *The Boatman*, *Wooden Boat* and *Classic Boat* cater for this market. In *Coasting* Raban muses more generally: 'It used to puzzle me that in every corner newsagent's in every English big city that I visited, there would be a stack of yachting magazines . . . Yachting-this and Yachting-that, the touched-up pornography of the wide spaces . . . Who bought these things?' (p38)

7 Raban, *The Oxford Book of the Sea*, p17. A 1923 meeting with Conrad which confirmed his conservative attitudes is described in David Bone, *Landfall at Sunset*, Duckworth (London) 1956, p157.

8 James Steele, *Queen Mary*, Phaidon Press Ltd (London) 1995.

9 See Wilson et al, *The Machine Age*, p147.

10 For a well-illustrated example see the conversion of the 1918 Dutch schooner *Grana* to the motor coaster *Capable* by FT Everard and Sons Ltd in 1925 (an auxiliary engine had been fitted in 1920, when the vessel was acquired). Details in KS Garrett, *Everard of Greenhithe*, The World Ship Society (Kendal) 1991, p65.

11 Jencks in Rose, p101.

12 Jencks, pp101, 102.

13 Jencks, p103.

14 Philip Dawson, 'Designing to the Bottom Line, a strategic approach to interior design by John McNeece for the business of passenger shipping', *Designs 95*, pp.104-110. Reference on p110.

15 Lane, p11 and Sekula, p51.

16 Robert Venturi, *Complexity and Contradiction in Architecture*, cited in Rose, p.106.

17 Rose, p102.

18 McNeece, p92.

19 Robinson and Herschman, *Architecture Transformed*, p112.

20 McNeece, p110.

21 Tillberg Design, working document for the *Queen of the Americas* project, 1995. Other unlocated quotations here are from this document.

22 Crawford in Sorkin, p3.

23 Crawford, p14.

24 Stephen Constantine, *Buy and Build: The Advertising Posters of the Empire Marketing Board*, Her Majesty's Stationery Office (London) 1986, p12.

25 Elleke Boehmer, *Colonial & Postcolonial Literature*, Oxford University Press (Oxford) 1995, p54.

26 EM Forster, *A Passage to India*, Penguin

(Harmondsworth) 1963 (first published 1924), p47.

27 Quoted in Fussell, p71.

28 Fussell, p171.

29 Anson, *Harbour Head*, pp1-2.

30 Sekula, p77.

31 Sekula, p12.

32 *Cruise & Ferry Interiors 1995/96*, p49.

33 See Meek, 'The First OCL Container Ships'.

34 Reyner Banham, *Theory and Design in the First Machine Age*, The Architectural Press (London) 1980 (first published 1960), p20.

35 Sudjic, p20.

36 Sudjic, p17

37 Sudjic, p51.

38 ibid.

39 Sudjic, p15.

40 Deyan Sudjic, *The 100 Mile City*, photographs by Phil Sayer, Andre Deutsch (London), p67

41 Eric de Maré, *The Nautical Style: An Aspect of the Functional Tradition*, Architectural Press Ltd (London), p8.

42 Virginia Woolf as quoted in Gavin Stamp's Introduction to Chris Ellmers and Alex Werner, *London's Lost Riverscape: A Photographic Panorama*, Viking (London) 1988, p3.

## CHAPTER 3

1 See Riccesi, especially pp66-142.

2 James Gardner, *The ARTful Designer: Ideas off the Drawing Board by James Gardner*, privately published (London) 1993, p287.

3 Gardner, p288.

4 Gardner, p296.

5 Gardner, p303.

6 Gardner 'No 736. [the builders' Yard Number for the *QE2*] Summary of Design Notes', dated 8 February 1967, at his London studio.

7 ibid.

8 Gardner, letter to Philip Dawson, 31 March 1988. Copy at Gardner's London studio.

9 Gardner, confidential letter to Commander Wood, August 1961. Copy at Gardner's London studio.

10 A letter to Gardner from Cunard dated 19 December 1961 expressed the need for the planned new vessel 'to maintain an Atlantic express service in the summer and cruise in the winter'. Copy at Gardner's London studio. For a detailed (if rather bland) account of the building and operation of the *QE2*, see Ronald Warwick and William Flayhart III, *QE2*, with a Foreword by HRH The Prince Philip, Duke of Edinburgh, WW Norton and Company (New York) 1985.

11 Gardner, 'The *QE2*, A Special Report', Supplement to *The Times*, 21 September 1967, pVI.

12 ibid.

13 ibid.

14 ibid.

15 ibid.

16 ibid.

17 Letter from Gardner to Philip Dawson, 31 March 1988: 'The ship, since Trafalgar Investments took over Cunard, has been worked over by amateurs, and [has] lost much of its coherence and style. Even the funnel has been clumsily modified for diesel outlets'. Copy at Gardner's London studio. For a detailed and well-illustrated account of general and specific design issues see Philip Dawson, *British Superliners of the Sixties: A Design Appreciation of the 'Oriana', 'Canberra' and 'QE2'*, Foreword by Sir Hugh Casson, Conway Maritime Press (London) 1990.

18 For a more detailed and entertaining account of the whole *QE2* design process see Gardner, *The ARTful Designer*, pp284-305, 'The birth of a white elephant'.

19 Dawson, p105.

20 The British naval architects I have interviewed in the course of research for this book hold this opinion.

21 Gardner, *The ARTful Designer*, p299.

22 Robert Hughes, *The Shock of the New*, British Broadcasting Corporation (London) 1980, p211.

23 For surveys of recent developments see the bibliography.

24 Sudjic, *New Directions in British Architecture*, p59.

25 Pye, *The Things We See: Ships*, p4.

26 Renzo Piano, 'Recupero del Porto di Genova per le Colombiadi', *Domus*, Milan, 740 July/August 1992, pp25-37. Reference on p30.

27 See Laurence Dunn's illustrated paper 'The Ship is Your Salesman – How Good is Her Image?' in J Anthony Hind (ed) *Ships' Gear*, Proceedings of the Technical Symposium on 'The Future Technical Patterns of Control Engineering, Equipment and Ship Operation' held in conjunction with the 'Ships' Gear International '66 Exhibition', Olympia, Fishing News (Books) Ltd (London) 1967, pp489-510. Dunn's writings on, and drawings of, ships of all kinds form a major resource for anyone seriously interested in the design of modern vessels.

28 Sudjic, *The 100 Mile City*, p158.

29 Ian Ritchie, 'Termini' in *Architecture of Transportation*, Architectural Design (London) 1994, p29.

30 Sekula, p45.

31 Sekula, p184.

32 Quoted Riccesi, p55.

33 Riccesi, p33.

34 Steele, p81; illustration on pp82-3.

35 Riccesi, p29.

36 ibid.

37 Riccesi, p72.

38 Gardner, letter to Philip Dawson, 31 March 1988.

39 op cit, Riccesi, p71; *Victoria* is illustrated on pp 70-1.

40 Rccesi, p30.

41 Philip Dawson, 'Your Own Patch, A Story of Balconies', *Guide 95*, pp107-12. Reference on p109.

42 Fussell, p41. Fussell has some acid perceptions on what he sees as the changing nature of travel: 'The calculated isolation from the actual which is tourism ("We fly you above the weather") is reflected as well in the design of the last of the serious liners, the QE2. Here the designers carefully eliminated the promenade deck, formerly the place where you were vouchsafed some proximity to the ocean . . . As the ship has been obliged to compete in the illusion of placelessness with the airport and the jet, its interior design has given over its former ambitions of alluding to such identifiable places as country estates . . . and instead has embraced the universal placeless style'. (p45)

43 Riccesi, p91.

44 Sekula, p113-14.

45 Lewis Mumford, *Technics and Civilization* (1934), as cited in Sekula, p114.

46 Giovanni Klaus Koenig, Introduction to Riccesi, p10.

47 Riccesi, p92.

48 Dunn, pp493 and 494. This ship was de-signed by Marshall Meek; it was his first large task and Dunn advised him on some aspects of the vessel's design. Today she is often to be seen anchored in Hong Kong harbour, and is still distinctive.

49 See *Designs 94* for a series of articles on the Hurtigruten vessels.

50 Lane, p52.

51 Lane, p188.

52 Lane, p189.

53 Mostert, p3.

54 Riccesi, p31.

55 Riccesi, p29.

56 Riccesi, p28.

57 Sudjic, *The 100 Mile City*, p6.

58 Sudjic, *The 100 Mile City*, p23.

59 Sudjic, *The 100 Mile City*, p25.

60 Sekula, p12.

61 Sekula, p49.

62 Mostert, p13.

63 Sudjic, *The 100 Mile City*, pp18 and 240.

64 Sudjic, *The 100 Mile City*, p97.

## CHAPTER 4

1 Walker, Fred, *The Song of the Clyde: A History of Clyde Shipbuilding*, WW Norton & Company (New York and London) 1984, p14.

2 See Waller, Chapter 4.

3 Kvaerner Masa-Yards has pioneered the installation of readymade and completely finished 'wet units' (bathrooms) and cabins in vessels.

4 David Tinsley, 'Compact coaster design proves a success', *Lloyd's List*, 24 January 1991, p4. I am grateful to Simon Leafe for this reference.

5 ibid.

6 ibid.

7 For a striking illustration of how ship repair work could literally overshadow residential property see the Frontispiece to John Falconer's *Sail and Steam: A Century of Seafaring Enterprise 1840-1935*, Viking (London) 1993, in which the barque *Penang* looms over Millwall back gardens.

8 Mostert, p135.

9 Steele, pp46-52.

10 Mostert, pp239-40.

11 Sekula, p101.

12 See John Lingwood, *SD14: The Great British Shipbuilding Success Story*, The World Ship Society (Kendal) 1976.

13 For related technology see Walker, pp190-91.

14 The most readable introduction to this rich field is Paul Fussell's *Abroad*. See also Terry Coleman, *The Liners: A History of the North Atlantic Crossing*, Penguin Books (Harmondsworth) 1977, first published Allen Lane 1976, and JM Brinnin, *The Sway of the Grand Saloon*, Delacorte (New York) 1971. Fussell comments: 'To travelers of the 20s like HM Tomlinson, the idea of travel is practically equivalent to the idea of ships, and the familiar spectacle of ships nearby . . . induced curious, subtle psychological ripples incident only to that time and place.

15 See Stephen M Payne, 'The Royal Yacht *Britannia*', *Ships Monthly* (Burton-on-Trent) Vol 30, No 10, October 1995, pp16-21.

16 See Laurence Dunn, *The World's Tankers*, Adlard Coles (London) 1956 for illustration and discussion of some examples.

17 See Wilkinson, *Wheelhouse and Bridge Design – A Shipbuilder's Appraisal* (1971), pp237-81.

18 Lane, p55. For an early account of designing accommodation for cargo vessels see Neville Ward, 'Some Aspects of the Interior Design of Ships', *Ships' Gear*, pp504-10 and Marshall Meek and Neville Ward, 'Accommodation in Ships', *Transactions of the Royal Institution of Naval Architects* (London), 1973, Vol 115, pp201-18.

19 Gardner, article in a Special Report on the *QE2*, *The Times*, 21 September 1967, pVI.

20 There are interesting analogies to be made between the high profile of certain car makes and their designers and the relatively anonymous styling of some of the best-loved models. Jaguar is a classic example of the latter.

21 I am grateful to William Everard for his time and expertise in explaining certain key issues.

22 Mostert, p89. See also Lane, Chapter VI 'Hierarchy'.

23 Dunn, 'The Ship is Your Salesman', p492.

24 Dunn, p489.

25 Dunn, p495.

26 See Gardiner (ed) *The Golden Age of Shipping*, p158.

27 Mostert, p145.

## CHAPTER 5

1 Paul Gilroy, *The Black Atlantic: Modernity and Double Consciousness*, Verso (London and New York) 1993.

2 Alan Hess comments in a 1995 e-mail to me: 'I would hazard a guess that they are influenced by the John Portman atria from the Hyatt hotel chain, an architectural form which has become a necessary cliche in any large hotel these days'.

3 I am indebted to Fred Walker for alerting me to this early example.

4 See Stephen Payne, 'The Evolution of Space: A History of the Atrium', *Designs 95* (Halmstad) 1995, pp115-18. The first multi-deck atrium was that on the 1987 Royal Caribbean Cruise Lines' *Sovereign of the Seas*; designed by Njärl Eide, it spanned five decks.

5 Tillberg Design project papers 1995.

6 'Market Report: A Matter of Safety?' *Guide 95* (Halmstad) 1995, p122.

7 For a recent report on this longstanding project see Nick Peters, 'Floating Town Liner Ready to Rule Waves', *The Sunday Times*, 10 December 1995, p17.

8 ibid.

9 'RCCL Atriums Open Up a Whole New World: How Njärl Eide used Ancient Greek Atriums to Revolutionise the Interiors of Modern Cruise Ships', *Cruise & Ferry Interiors 1995/96*, pp33-4.

10 Mike Davis, *City of Quartz: Excavating the Future in Los Angeles*, Vintage (London) 1992. Davis comments of 'Fortress LA': 'As a prestige symbol – and sometimes as the decisive borderline between the merely well-off and the "truly rich" – "security" has less to do with personal safety than with the degree of personal insulation, in residential, work, consumption and travel environments, from "unsavory" groups and individuals, even crowds in general'. (p224)

11 Charles Jencks, *The New Moderns, From Late to Neo-Modernism*, Academy Editions (London) 1990, p37.

12 Toumas Routa, 'Application of the SWATH principle to passenger vessels', *Proceedings* of International Conference on SWATH and Advanced Multi-Hulled Vessels held at The Royal Institution of Naval Architects, London, 17-19 April, 1985, Section 3.2.

13 Routa, Section 5.3.

14 K Levander, discussion after Routa paper, Section 5.4.

15 Jencks, p98.

# SELECTED BIBLIOGRAPHY

This bibliography is selective on the basis of general interest and the quality of illustrations; the kind I wish I had encountered years ago. The best general magazine in Britain on modern commercial ships and their design is *Ships Monthly*; earlier journals, along with the invaluable Transactions of the Institute of Marine Engineers and of the Royal Institution of Naval Architects, can be found only in specialist libraries. Of interest are runs of *Architectural Review, Art and Industry, Casabella, Design, Domus, Industrial Design, Journal of Shipping and Commerce, The Motorship, The Naval Architect, Shipbuilding and Shipping Record, The Shipping World, Syren and Shipping Illustrated, The Studio, The Shipbuilder, Shipbuilding and Shipping Record, Shipping World and Shipbuilder, Marine Engineer and Naval Architect.*

## BOOKS

– Al Naib, SK (ed), *Dockland: An illustrated Historical Survey of Life and Work in East London*, North East London Polytechnic in conjunction with the Greater London Council (London) 1986.

– Anson, Peter F, *Harbour Head: Maritime Memories*, The Catholic Book Club (London) 1944; *Fishing Boat and Fisher Folk on the East Coast of Scotland*, with an Introduction by Alexander Fenton, JM Dent and Sons Ltd (London) 1930 (reprinted Dent 1971); *How to Draw Ships*, The Studio Publications (London and New York) 1955 (first published 1941).

– Baker, WA, *From Paddle Steamer to Nuclear Ship: A History of the Engine-Powered Vessel*, CA Watts & Co Ltd (London) 1965.

– Bathe, Basil W, *Seven Centuries of Sea Travel, From the Crusaders to the Cruises*, Portland House (New York) 1973.

– Baynes, Ken and Pugh, Francis, *The Art of the Engineer*, Lund Humphries Publishers Ltd (London) 1981.

– Bel Geddes, Norman, *Horizons*, Little, Brown & Co (New York) 1932.

– Bent, Mike, *Steamers of the Fjords: Bergen Shipping since 1839*, Conway Maritime Press (London) 1989.

– Bertram, Anthony, *Design*, Penguin (Harmondsworth) 1938.

– Biesty, Stephen (illustrations) and Richard Platt (text), *Stephen Biesty's Incredible Cross-Sections*, Dorling Kinderlsey (London, New York, Stuttgart) 1992.

– Constantine, Stephen, *Buy & Build: The Advertising Posters of the Empire Marketing Board*, Her Majesty's Stationery Office (London) 1986.

– Corbin, Alain, *The Lure of the Sea: The Discovery of the Seaside in the Western World 1750-1840*, Penguin Books (Harmondsworth) 1994 (first published as *Le Territoire Vide*, Paris, 1988).

– Corlett, Ewan, *The Ship: The Revolution in Merchant Shipping 1950-1980*, National Maritime Museum/Her Majesty's Stationery Office (London) 1981 (The final volume in a 10-volume series).

– Craig, Robin, *The Ship: Steam Tramps and Cargo Liners 1850-1950*, National Maritime Museum/Her Majesty's Stationery Office (London) 1980 (Volume 5 in a 10-volume series).

– Cuyvers, Luc, *Sea Power: A Global Journey*, Naval Institute Press (Annapolis, Maryland) 1993.

– Davis, Mike, *City of Quartz: Excavating the Future in Los Angeles*, photographs by Robert Morrow, Vintage (London) 1992.

– Dawson, Philip, *British Superliners of the Sixties: A Design Appreciation of the 'Oriana', 'Canberra' and 'QE2'*, Foreword by Sir Hugh Casson, Conway Maritime Press (London) 1990.

– De Maré, Eric, *The Nautical Style: An Aspect of the Functional Tradition*, Architectural Press (London) 1973.

– Dunn, Laurence, *The World's Tankers*, Adlard Coles Ltd in association with George G Harrap and Company Ltd (London) 1956; *Famous Liners of the Past: Belfast Built*, written and illustrated by Laurence Dunn, Adlard Coles (London) 1964; *Passenger Liners*, Foreword by CM Squarey, Adlard Coles Ltd (London) 1961, revised edition 1965; *Merchant Ships of the World in Colour, 1910-1929*, Blandford Press (London) 1973; *Laurence Dunn's Thames Shipping*, Carmania Press (London) 1992.

– Ellmers, Chris and Werner, Alex, *London's Lost Riverscape: A Photographic Panorama*, Introduction by Gavin Stamp, Viking (London) 1988; *Dockland Life: A Pictorial History of London's Docks 1860-1970*, Museum of London/Mainstream Publishing (Edinburgh and London) 1991.

– Falconer, John, *Sail and Steam: A Century of Seafaring Enterprise 1840-1935*, photographs from The National Maritime Museum, Greenwich, Viking (London) 1993.

– Fishman, William J, Breach, Nicholas and Hall, John M, *East End and Docklands*, Duckworth (London) 1990.

– Fowles, John (text) and the Gibsons of Scilly (photographs) *Shipwreck*, Jonathan Cape (London) 1974.

– Frere-Cook, Gervis, *The Decorative Arts of the Mariner*, Jupiter Books (London) 1974. First published Cassell (London) 1966.

– Fricker, Philip J, *Becken of Cowes: Ocean Liners*, Reed (London) 1992.

– Frost, Ted, *From Tree to Sea: The Building of a Wooden Steam Drifter*, Terence Dalton Ltd (Lavenham) 1985.

– Fussell, Paul, *Abroad: British Literary Traveling Between the Wars*, Oxford University Press (New York and Oxford) 1980.

– Gardner, James, *The ARTful Designer: Ideas off the Drawing Board by James Gardner*, privately printed (London) 1993.

– Gardiner, Robert (ed), *The Shipping Revolution: The Modern Merchant Ship*, Conway Maritime Press (London) 1992; (ed) *The Golden Age of Shipping: The Classic Merchant Ship 1900-1960*, Conway Maritime Press (London) 1994.

– Gaston, MJ, *Tugs and Towing*, Patrick Stephens/Haynes Publishing (Sparkford) 1991.

– Gillmer, Thomas C, *Modern Ship Design*, second edition, Naval Institute Press (Annapolis, Maryland) 1975.

– Groenendijk, Paul (text), Linders, Jannes and Rook, Piet (photographs), *Scheepsarchitectuur* (Uitgeverij OIO, Rotterdam) 1989 (Dutch text).

– Guiton, J, *Aesthetic Aspects of Ship and Yacht Design*, Adlard Coles (London) 1971.

– Harlaftis, Gelina, *Greek Seamen and Greek Steamships on the Eve of the First World War*, Aegean Maritime Museum (Myconos) 1994.

– Harris, CJ & Ingpen, Brian D, *Mailships of the Union-Castle Line*, Patrick Stephens/Haynes Publishing (Sparkford) 1994 (originally published in South Africa by Fernwood Press, 1994.)

– Harris, Daniel G, Chapman, FH, *The First Naval Architect and His Work*, Conway Maritime Press (London) 1989.

– Haws, Duncan, *Ships and the Sea: A Chronological Review*, Hart Davis, MacGibbon (London) 1976.

– Heskett, John, *Industrial Design*, Thames and Hudson (London) 1980; *Design in Germany 1870-1918*, Trefoil Design Library (London) 1986.

– Hess, Alan, *Viva Las Vegas! After Hours Architecture*, Foreword by Robert Venturi, Denise Scott Brown and Steven Izenour, Chronicle Books (San Francisco) 1993.

– Hillion, Daniel, *La Mère s'Affiche*, Editions Ouest (Rennes) 1990 (French text).

– Hughes, Quentin, *Architecture & Townscape in Liverpool*, with photographs by Graham Smith and David Wrightman, Lund Humphries (London) 1964.

– Ingpen, Brian and Pabst, Robert, *Maritime South Africa: A Pictorial History*, Jane's Publishing Company Ltd (London) 1984.

– Ireland, Bernard, *The Busy Channel*, Ian Allan (London) 1981.

– Jackson, Gordon, *The History and Archeology of Ports*, World's Work Ltd/The Windmill Press (Tadworth) 1983.

– Jencks, Charles, *The Language of Post-Modern Architecture*, Academy Editions (London), sixth revised enlarged edition, 1991.

– Johnston, Ian, *Beardmore Built: The Rise and Fall of a Clydeside Shipyard*, Clydebank District Libraries & Museums Department (Clydebank) 1993.

– Kentley, Eric, *Boat: Discover the Story of Boats and Ships – from Dugout Canoes to Luxury Yachts and Giant Supertankers*, Dorling Kindersley (London, New York, Stuttgart) in association with The National Maritime Museum (London) 1992.

– Klingender, Francis D, *Art and the Industrial Revolution*, edited and revised by Arthur Elton, Paladin (St Albans) 1972.

– Krieger, Michael J, *Tramp: Sagas of High Adventure in the Vanishing World of the Old Tramp Freighters*, Photography by Judy Howard, Angus & Robertson UK (London) 1987. First published by Chronicle Books (San Francisco) 1986.

– Kronenberg, Robert, *Houses in Motion: The Genesis, History and Development of the Portable Building*, Academy Editions (London) 1995.

– Kummerman, Henri and Jacquinet, Robert (eds), *Ships' Cargo, Cargo Ships*, MacGregor Publications Ltd (London) 1979.

– Landström, Björn, *The Ship: A Survey of the History of the Ship from the Primitive Raft to the Nuclear-powered Submarine with Reconstructions in Words and Pictures*, Allen & Unwin (London) 1961 (originally published in Sweden as *Skeppet*, 1961).

– Lane, Tony, *Grey Dawn Breaking: British Merchant Seafarers in the Late Twentieth Century*, Manchester University Press (Manchester) 1986.

– Lingwood, J, *Significant Ships of 1990* [then annually] (Royal Institution of Naval Architects, London) 1990.

– Lord, John (text) and Southam, Jem (photographs). *The Floating Harbour: A Landscape History of Bristol City Docks*, The Redcliffe Press (Bristol) 1983.

– Loewy, Raymond, *Industrial Design*, (London and New York) 1979.

– Lundbladh, Janne (ed), *The Lore of Ships*, AB Nordbok (Gothenburg) 1975.

– Lynch, Don (text) and Marschall, Ken (paintings), *Titanic, An Illustrated History*, Introduction by Robert D Ballard, Hodder & Stoughton (London) 1992.

– Maber, John, *The Ship: Channel Packets and Ocean Liners 1850-1970*, National Maritime Museum/Her Majesty's Stationery Office (London) 1980 (Volume 6 in a ten-volume series).

– Marshall, Chris, *The Encyclopedia of Ships: the History and Specifications of Over 1200 Ships*, Blitz Editions/Orbis (Leicester) 1995 (first published Istituto Geografico De Agostini, SpA, Novara 1995).

– McCaughan, Michael, *Steel Ships & Iron Men: Ship-*

building in Belfast, 1894-1912, The Harland & Wolff Historic Photograph Collection at the Ulster Folk & Transport Museum, The Friar's Bush Press (Belfast) 1989.

– MacRae, JA and Waine, CV, The Steam Collier Fleets, Illustrated by CV Waine, Waine Research Publications (Albrighton) 1990.

– McKean, John, Crystal Palace: Joseph Paxton and Charles Fox, Phaidon Press Ltd (London) 1994.

– McKee, Eric, Working Boats of Britain: Their Shape and Purpose, Introduction by Dr Basil Greenhill, Conway Maritime Press in association with the National Maritime Museum (London) 1983.

– Moore, David, Sydney Harbour, text by Rodney Hall, Chapter & Verse in association with State Library of New South Wales Press (Sydney) 1993.

– Mostert, Nöel, Supership, Macmillan (London) 1974.

– Munro-Smith, R, Merchant Ship Types, Marine Media Management Ltd for Royal Institution of Naval Architects (London), 1975.

– Plowden, David, End of An Era: The Last of the Great Lakes Steamboats, WW Norton & Company (New York and London) 1992.

– Plowman, Peter, Emigrant Ships to Luxury Liners: Passenger Ships to Australia and New Zealand 1945-1990, New South Wales University Press (Kensington, NSW) 1992.

– Pye, David, The Things We See: Ships, Penguin Books (Harmondsworth) 1950; The Nature and Art of Workmanship, Cambridge University Press (Cambridge) 1968; The Nature and Aesthetics of Design, Barrie & Jenkins (London) 1978.

– Quarm, Roger and Wilcox, Scott, Masters of the Sea: British Marine Watercolours, Phaidon Press Ltd (Oxford) in association with the National Maritime Museum, Greenwich, and the Yale Center for British Art (New Haven, Connecticut) 1987.

– Raban, Jonathan, Coasting, Collins Harvill (London) 1986; (ed) The Oxford Book of The Sea, Oxford University Press (Oxford and New York) 1992.

– Ransome-Wallis, P, Merchant Ship Panorama, Ian Allan Ltd (London) 1980. (The Ransome-Wallis Collection of ship photographs is now held at the National Maritime Museum.)

– Ratcliffe, Mike, Liquid Gold Ships: A History of the Tanker, Lloyd's of London Press (London) 1985.

– Reid, George, River Thames in the Late Twenties & Early Thirties, edited by Mike Seaborne, text by Chris Ellmers, Dirk Nishen Publishing (London) 1987.

– Riccesi, Donato, Gustavo Pulitzer Finali: Il Disegno della Nave, Allestimento Interni 1925-1967, Marsilio Editori (Venice) 1985 (Italian text).

– Ritchie-Noakes, Nancy, Liverpool's Historic Waterfront: The World's First Mercantile Dock System, Her Majesty's Stationery Office (London) 1984.

– Roiter, Fulvio (photographs) Marco Ferranti (text) Crown Princess, Vianello Libri (Treviso)/Fincantieri (Triest) 1990.

– Rose, Margaret A, The Post-Modern and the Post-Industrial: A Critical Analysis, Cambridge University Press (Cambridge) 1991.

– Salgado, Sebastião, Workers: An Anthology of the Industrial Age, Phaidon Press Ltd (London) 1993.

Schaap, Dick, A Bridge to the Seven Seas: A Hundred Years of the Holland America Line, Patrick Stephens (London) 1973.

– Scheidulin, Hans and Gutmann, Hermann, The Ports: Bremen, Bremerhaven, HH Hauschild GmbH (Bremen) 1985.

– Schönknecht, Rolf, Lüsch, Jürgen, Schelzel, Manfred and Obenaus, Hans, Ships and Shipping of Tomorrow, MacGregor Publications Ltd, VEB Verlag Technik (Berlin) 1983 (originally published in German, 1973)

Sekula, Allan, Fish Story, Witte de Wish, Center for Contemporary Art (Rotterdam) and Richter Verlag (Düsseldorf) 1995.

– Smith, B, Merchant Ship Design since 1945, Ian Allan Ltd (London) 1984.

– Sorkin, Michael, Variations on a Theme Park: The New American City and the End of Public Space, Hill and Wang (New York) 1992.

– Steele, James, Queen Mary, Phaidon Press Ltd (London) 1995.

– Stewart, Peter, The Beauty of Ships, with an Introductory Essay by Peter Stewart, Max Parrish (London) 1963.

– Sudjic, Deyan, The 100 Mile City, photographs by Phil Sayer, Andre Deutsch (London) 1992.

– Teague, Walter Dorwin, Design This Day: The Technique of Order in the Machine Age, The Studio Publications (New York 1940 and London 1947).

– Tinsley, David, Significant Ships of 1994, Royal Institution of Naval Architects (London) 1995 (an annual publication).

– Tritton, Roger (ed) The Visual Dictionary of Ships and Sailing, Dorling Kindersley (London) 1991.

– Venturi, Robert, Scott Brown, Denise, Izenour, Steven, Learning from Las Vegas: The Forgotten Symbolism of Architectural Form, The MIT Press (Cambridge, Massachusetts and London), thirteenth printing 1994, first published 1972.

– Waine, Charles V, Steam Coasters and Short Sea Traders, illustrated by the author, Waine Research Publications (Albrighton) 1980.

– Walker, David (text) and Barrett, Wayne (photographs), Small Wooden Boats of the Atlantic, Introduction by Marven E Moore, Nimbus Publishing Ltd (Halifax, Nova Scotia) 1990.

– Walker, Fred M, Song of the Clyde: A History of Clyde Shipbuilding, WW Norton & Company (New York and London) 1984.

– Wenzel, Heinz (ed) The Sea: The Largest Continent in the World, Kummerley & Frey (Berne) 1961; The Eternal Sea, A Pictorial Anthology, translated by Walter Kaufman, Abelard-Schuman (London, Toronto and New York) 1970.

– Wilson, Cliff (photographs) Christopher Rush (text), Where the Clock Stands Still: A Portrait of the East Neuk of Fife, Canongate (Edinburgh) 1990.

– Wilson, Richard Guy, Pilgrim, Dianne H, Tashjian, Dickran, The Machine Age in America, The Brooklyn Museum in association with Harry N Abrams, Inc (New York) 1986.

– Winchester, Clarence (ed), Shipping Wonders of the World: A Saga of the Sea in Story and Picture, 2 vols, The Amalgamated Press Ltd (London) [nd, 1937]

JOURNALS AND CATALOGUES

– Brogren, Klas, Designs 95, Plus 2 Ferryconsultation (Halmstad) 1995 (an annual publication); Guide 95, Plus 2 Ferryconsultation (Halmstad) 1995 (an annual publication).

– Euromoney Publications (London) 1995, Cruise Ship & Ferry Interiors 1995/96.

– National Maritime Museum (London) 1990, The National Maritime Museum: The Collections.

– Stonham, Denis and National Maritime Museum Staff, Historic Photographs at the National Maritime Museum: An Illustrated Guide, Ashford, Buchan and Enright (Leatherhead) 1995.

# INDEX

# ACKNOWLEDEGEMENTS

The researching and writing of this book has been a haven in a storm; my acknowledgements here are no formality. The 1990 exhibition and catalogue 'Scheepsarchitectuur' by Albert Roskam of Stichting Kunstprojecten, Paul Groenendijk, Jannes Linders and Piet Rook was my direct inspiration. Gavin Stamp at the Mackintosh School of Architecture, University of Glasgow and the Glasgow School of Art provided invaluable early contacts with Caroline Larkin at The Architecture Foundation, and Kate Trant at the Royal Institute of British Architects was also most supportive. Staff at the London libraries of the Architectural Association, The Institute of Marine Engineers, and of the Royal Institution of Naval Architects, were unfailingly helpful.

Without both the imaginative and practical support of the National Maritime Museum, Greenwich, this book could never have happened and Roger Knight gave early encouragement. I thank especially David Spence, Margarette Lincoln, Karen Peart and Stephen Riley, and in the archives at Woolwich Arsenal Bob Todd and David Hodge. The fine photographs here by Tina Chambers and Peter Robinson make this book, in a very real sense, theirs as well. At Academy Editions I am grateful to Andrea Betella and to Alex Young for skilled design work, to Ramona Khambatta for her efficient editing, and to Maggie Toy for first trusting me; together they made work a pleasure.

The generous cooperation of the port authorities of Dover, Felixstowe, Harwich, London, Tilbury and Southampton enabled many of the contemporary photographs in this book to be specially taken. The following individuals and firms were also outstandingly generous with their time and resources in helping me: in Australia the photographer David Moore, Sydney; in Britain Costa Cruises, Benetton Group, John Batram at P&O Cruises, Sir Hugh Casson, Professor William Couper at the University of Cardiff, William Everard, Director of FT Everard and Sons Ltd, Greenhithe, and also the Company's Marine Director, Captain KS Garrett, the Evergreen International Corporation Ltd, Eve Harrison and the staff at James Gardner's Studio, London, Ian Johnston at the Glasgow School of Art and Helena Frith-Powell at Euromoney Publications, London, gave much-appreciated help, Simon Leafe, Director of Leafe & Hawkes Ltd, Hull (who also got me to sea again!), Jim Lindsay at The Scottish Fisheries Museum Trust, Anstruther, John McLean at the University of Brighton, Stephen Rabson at P&O Head Office, the photographer Dennis Stonham, Matthew Tanner at Merseyside Maritime Museum, Jim Wilson, Managing Director of Appledore Shipbuilders Ltd; in Finland Henrik Segercrantz at Kvaerner Masa-Yards, Helsinki; in Germany Captain Wilfried Hammann of Hammann & Prahm, Leer; in Italy Gianfranco Bertaglia of Fincantieri, Triest; in the Netherlands the photographer Piet Rook; in Norway Bill Bland, Managing Director of Skipskonsulent AS, Laksevaag; in Sweden Klas Brogren of Plus 2 Ferryconsultation, Halmstad, the photographer Johan Garsten, Bromma and Robert and Tomas Tillberg of Tillberg Design, Viken; in the United States Joseph Farcus PA, Miami Beach, architect of floating worlds, John Heskett at Illinois Institute of Design, the photographer and (intellectual) maritime explorer Allan Sekula in Los Angeles and the architect Alan Hess in Rancho Santa Margarita, California, inspired historian of the everyday environment.

Special thanks go to Ian Ritchie of Ian Ritchie Architects, London, for his time in explaining architectural contexts, and to Marshall Meek CBE, Gianfranco Bertaglia and Fred Walker for insights into the theory and practice of naval architecture; Fred also read extensive early drafts of the book, and furnished invaluable advice.

I gratefully acknowledge financial support from the National Maritime Museum's Caird Fund, from AmCAS and from the Research Committee of the School of English and American Studies at the University of Exeter. John Saunders and Tony Fisher of the University of Exeter's Photographic Unit provided enthusiastic and skilled technical support, and Katherine Fenton saved my computer bacon.

Early versions of some chapters here were given as research papers during 1995 to the Staff Research Group at the National Maritime Museum, Greenwich, to the Research Seminar of the School of English and American Studies and to the second annual conference of the Design History Society, held at the University of Plymouth.

I am indebted to Ron Tamplin for memorable imaginative voyages in the South Seas, and to Chris Brooks for sympathetic architectural perceptions, while Peter New was a supportive Head of School. Tony Clayden, Associate Dean of the Faculty of Arts and Education, University of Plymouth, gave more encouragement than he could ever guess. Finally, without the attitudes fostered by Keith Joseph and Margaret Thatcher in the wake of the Falklands War I might never have turned again – for sanity – to ships and the sea: 'Rejoice!'

Despite all advice received, this thing is mine alone and must fare accordingly.
PQ
Exeter
Commonwealth Day 1996

# COPYRIGHT